M000104904

Algebra I

PRACTICE
MAKES
PERFECT™

Algebra I
Review and Workbook

SECOND EDITION

Carolyn Wheater

New York   Chicago   San Francisco   Athens   London   Madrid   Mexico City
Milan   New Delhi   Singapore   Sydney   Toronto

5 6 7 8 9   LHS   22 21 20

ISBN   978-1-260-02644-3
MHID   1-260-02644-2

e-ISBN   978-1-260-02645-0
e-MHID   1-260-02645-0

McGraw-Hill Education books are available at special quantity discounts to use as premiums and sales promotions or for use in corporate training programs. To contact a representative, please visit the Contact Us pages at www.mhprofessional.com.

Contents

5 Systems of linear equations and inequalities 37

6 Powers and polynomials 47

7 Factoring 59

8 Radicals 69

Introduction

An old joke tells of a tourist, lost in New York City, who stops a passerby to ask, "How do I get to Carnegie Hall?" The New Yorker's answer comes back quickly: "Practice, practice, practice!" The joke may be lame, but it contains a truth. No musician performs on the stage of a renowned concert hall without years of daily and diligent practice. No dancer steps out on stage without hours in the rehearsal hall, and no athlete takes to the field or the court without investing time and sweat drilling on the skills of his or her sport.

Math has a lot in common with music, dance, and sports. There are skills to be learned and a sequence of activities you need to go through if you want to be good at it. You don't just read math, or just listen to math, or even just understand math. You *do* math, and to learn to do it well, you have to practice. That's why homework exists, but most people need more practice than homework provides. That's where *Practice Makes Perfect Algebra* comes in.

When you start your formal study of algebra, you take your first step into the world of advanced mathematics. One of your principal tasks is to build the repertoire of tools that you will use in all future math courses and many other courses as well. To do that, you first need to understand each tool and how to use it, and then how to use the various tools in your toolbox in combination.

The almost 1000 exercises in this book are designed to help you acquire the skills you need, practice each one individually until you have confidence in it, and then combine various skills to solve more complicated problems. Since it's also important to keep your tools in good condition, you can use *Practice Makes Perfect Algebra* to review. Reminding yourself of the tools in your toolbox and how to use them helps prepare you to face new tasks that require you to combine those tools in new ways.

One tool that is relatively new but growing in importance is the calculator, and specifically, the graphing calculator. Generations of students learned algebra without using any sort of calculator, and if you do not have access to one, you can still learn all the algebra you need. As calculators became available, they provided the opportunity to explore ideas without worrying about whether the arithmetic would get too difficult. The rise of graphing calculators means that you can investigate properties of functions and their graphs without spending lots of time drawing those graphs by hand.

Throughout this book, you'll see Calculator Notes. These are ideas on how a graphing calculator might help you check your work, or solve a problem when you're stuck. These tips are not meant to replace you learning the skills and doing the work. That will always be essential. The Notes are based on a commonly used graphing calculator, which is sometimes introduced in algebra. If you have one, you might be interested in the Notes. If you don't have one, there's no need to rush to get one.

With patience and practice, you'll find that you've assembled an impressive set of tools and that you're confident about your ability to use them properly. The skills you acquire in algebra will serve you well in other math courses and in other disciplines. Be persistent. You must keep working at it, bit by bit. Be patient. You will make mistakes, but mistakes are one of the ways we learn, so welcome your mistakes. They'll decrease as you practice, because practice makes perfect.

PRACTICE
MAKES
PERFECT™

Algebra I

Arithmetic to algebra

In arithmetic, we learn to work with numbers: adding, subtracting, multiplying, and dividing. Algebra builds on that work, extends it, and reverses it. Algebra looks at the properties of numbers and number systems, introduces the use of symbols called variables to stand for numbers that are unknown or changeable, and develops techniques for finding those unknowns.

The real numbers

The real numbers include all the numbers you encounter in arithmetic. The natural, or counting, numbers are the numbers you used as you learned to count: {1, 2, 3, 4, 5, …}. Add the number 0 to the natural numbers and you have the whole numbers: {0, 1, 2, 3, 4, …}. The whole numbers together with their opposites form the integers, the positive and negative whole numbers and 0: {…, −3, −2, −1, 0, 1, 2, 3, …}.

There are many numbers between each pair of adjacent integers, however. Some of these, called rational numbers, are numbers that can be expressed as the ratio of two integers, that is, as a fraction. All integers are rational, since every integer can be written as a fraction by giving it a denominator of 1. Rational numbers have decimal expansions that either terminate (like $\frac{5}{2} = 2.5$) or infinitely repeat a pattern (like $\frac{1}{3} = 0.33333\ldots = 0.\overline{3}$).

There are still other numbers that cannot be expressed as the ratio of two integers, called irrational numbers. These include numbers like π and $\sqrt{2}$. You may have used decimals to approximate these, but irrational numbers have decimal representations that continue forever and do not repeat. For an exact answer, leave numbers in terms of π or in simplest radical form. When you try to express irrational numbers in decimal form, you're forced to cut the infinite decimal off, and that means your answer is approximate.

The real numbers include both the rationals and the irrationals. The number line gives a visual representation of the real numbers (see Figure 1.1). Each point on the line corresponds to a real number.

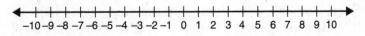

Figure 1.1 The real number line.

For each number given, list the sets of numbers into which the number fits (naturals, wholes, integers, rationals, irrationals, or reals).

1. 17.386

2. −5

3. $\dfrac{2}{5}$

4. 0

5. $\sqrt{7}$

6. 493

7. −17.5

8. $73.\overline{874}$

9. $\dfrac{123}{-41}$

10. π

For 11-20, plot the numbers on the real number line and label the point with the appropriate letter. Use the following figure for your reference.

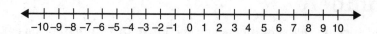

11. $A = 4.5$

12. $B = \sqrt{10}$

13. $C = -2.75$

14. $D = 0.1$

15. $E = \dfrac{25}{4}$

16. $F = -\sqrt{20}$

17. $G = \dfrac{-46}{5}$

18. $H = 7.25$

19. $I = -6\dfrac{1}{3}$

20. $J = 8.9$

Properties of real numbers

As you learned arithmetic, you also learned certain rules about the way numbers behave that helped you do your work more efficiently. You might not have stopped to put names to those properties, but you knew, for example, that $4 + 5$ was the same as $5 + 4$, but $5 − 4$ did not equal $4 − 5$.

The commutative and associative properties are the rules that tell you how you can rearrange the numbers in an arithmetic problem to make the calculation easier. The commutative property tells you when you may change the order, and the associative property tells you when you can regroup. There are commutative and associative properties for addition and for multiplication.

Commutative Property for Addition: $a + b = b + a$ [Example: $5 + 4 = 4 + 5$]
Commutative Property for Multiplication: $a \times b = b \times a$ [Example: $3 \times 5 = 5 \times 3$]
Associative Property for Addition: $(a + b) + c = a + (b + c)$ [Example: $(3 + 4) + 5 = 3 + (4 + 5)$]
Associative Property for Multiplication: $(a \times b) \times c = a \times (b \times c)$ [Example: $(2 \times 3) \times 4 = 2 \times (3 \times 4)$]

Two other properties of the real numbers sound obvious, but we'd be lost without them. The identity properties for addition and multiplication say that there is a real number—0 for addition and 1 for multiplication—that doesn't change anything. When you add 0 to a number or multiply a number by 1, you end up with the same number.

Identity for Addition: $a + 0 = a$
Identity for Multiplication: $a \times 1 = a, a \neq 0$

The inverse properties guarantee that whatever number you start with, you can find a number to add to it, or to multiply it by, to get back to the identity.

Inverse for Addition: $a + -a = 0$ [Example: $4 + -4 = 0$]

Inverse for Multiplication: $a \times \dfrac{1}{a} = 1, a \neq 0$ [Example: $2 \times \dfrac{1}{2} = 1$]

Notice that 0 doesn't have an inverse for multiplication. That's because of another property you know but don't often think about. Any number multiplied by 0 equals 0.

Multiplicative Property of Zero: $a \times 0 = 0$

It's interesting that while multiplying by 0 always gives you 0, there's no way to get a product of 0 without using 0 as one of your factors.

Zero Product Property: If $a \times b = 0$, then $a = 0$ or $b = 0$ or both.

Finally, the distributive property ties together addition and multiplication. The distributive property for multiplication over addition—its full name—says that you can do the problem in two different orders and get the same answer. If you want to multiply $5 \times 40 + 8$, you can add $40 + 8 = 48$ and then multiply 5×48, or you can multiply $5 \times 40 = 200$ and $5 \times 8 = 40$, and then add $200 + 40$. You get 240 either way.

Distributive Property: $a(b + c) = a \times b + a \times c$

EXERCISE

1·2

Identify the property of the real number system that is represented in each example.

1. $7 + 6 + 3 = 7 + 3 + 6$

2. $(5 \times 8) \times 2 = 5 \times (8 \times 2)$

3. $4 + 0 = 4$

4. $2 \times \dfrac{1}{2} = 1$

5. $8(3 + 9) = 8 \times 3 + 8 \times 9$

6. $5x = 0$, so $x = 0$

7. $(8 + 3) + 6 = 8 + (3 + 6)$

8. $28 \times 1 = 28$

9. $7 \times 4 \times 9 = 4 \times 7 \times 9$

10. $193 \times 0 = 0$

11. $14 + (-14) = 0$

12. $3(58) = 3 \times 50 + 3 \times 8$

13. $\dfrac{2}{3} \times \dfrac{6}{6} = \dfrac{12}{18}$

14. $(4 + 1) + 9 = 4 + (1 + 9)$

15. $839 + (-839) = 0$

Integers

The integers are the positive and negative whole numbers and 0. On the number line, the negative numbers are a mirror image of the positive numbers; this can be confusing sometimes when you're thinking about the relative size of numbers. On the positive side, 7 is larger than 4, but on the negative side, −7 is less than −4. It may help to picture the number line and think about "larger" as farther right and "smaller" as farther left.

Expanding your understanding of arithmetic to include the integers is a first big step in algebra. When you first learned to subtract, you would have said you couldn't subtract 8 from 3, but when you open up your thinking to include negative numbers, you can. The rules for operating with integers apply to all real numbers, so it's important to learn them well.

Absolute value

The absolute value of a number is its distance from 0 without regard to direction. If a number and its opposite are the same distance from 0, in opposite directions, they have the same absolute values. $|4|$ and $|-4|$ both equal 4, because both 4 and −4 are four units from 0.

Addition

To add integers with the same sign, add the absolute values and keep the sign. Add $4 + 7$, both positive numbers, and you get 11, a positive number. Add $-5 + (-3)$, both negative, and you get −8.

To add integers with different signs, subtract the absolute values and take the sign of the integer with the larger absolute value. If you need to add $13 + (-5)$, think $13 - 5 = 8$, then look back and see that the larger-looking number, 13, is positive, so your answer is positive 8. On the other hand, $9 + (-12)$, is going to turn out negative because $|-12| > |9|$. You'll wind up with −3.

Subtraction

Did you notice that none of the properties of the real numbers talked about subtraction? That's because subtraction is defined as addition of the inverse. To subtract 4, you add −4; to subtract −9, you add 9. When you learned to subtract, to answer questions like $8 - 5 = \boxed{}$, what you were really doing was answering $\boxed{} + 5 = 8$. Every subtraction problem is an addition problem in disguise.

To subtract an integer, add its opposite. Change the sign of the second number (the *subtrahend*, if you want the mathematical term) and add. The problem $9 - (-7)$ becomes $9 + 7$, whereas $-3 - 8$ becomes $-3 + (-8)$. Then you follow the rules for addition.

$$9 - (-7) = 9 + 7 = 16$$
$$-3 - 8 = -3 + (-8) = -11$$

Multiplication

To multiply two integers, multiply the absolute values and then determine the sign. If the integers have the same sign, the product will be positive. If the factors have different signs, the product is negative.

$$4 \times 7 = 28$$
$$-4 \times (-7) = 28$$
$$4 \times (-7) = -28$$
$$-4 \times 7 = -28$$

Division

Just as subtraction is defined as adding the inverse, division is defined as multiplying by the inverse. The multiplicative inverse, or reciprocal, of an integer n is $\frac{1}{n}$. To form the reciprocal of a fraction, swap the numerator and denominator. The reciprocal of 4 is $\frac{1}{4}$ and the reciprocal of $\frac{3}{4}$ is $\frac{4}{3}$. You probably remember learning that to divide by a fraction, you should invert the divisor and multiply.

Since division is multiplication in disguise, you follow the same rules for signs when you divide that you follow when multiplying. To divide two integers, divide the absolute values. If the signs are the same, the quotient is positive. If the signs are different, the quotient is negative.

EXERCISE

1·3

Find the value of each expression.

1. $-12 + 14$
2. $-13 - 4$
3. $18 \times (-3)$
4. $-32 \div (-8)$
5. $6 + (-3)$
6. $5 - (-9)$
7. 2×12
8. $12 \div (-4)$
9. $-6 - 2$
10. $-9 \times (-2)$

11. $-5 + 7$
12. $12 - 5$
13. $8 \times (-4)$
14. $-2 - 8$
15. -5×8
16. $-9 - 3$
17. $5 \div (-5)$
18. -4×12
19. $-4 \times (-4)$
20. $-45 \div (-9)$

Order of operations

The order of operations is an established system for determining which operations to perform first when evaluating an expression. The order of operations tells you first to evaluate any expressions in parentheses. Exponents are next in the order, and then, moving from left to right, perform any multiplications or divisions as you meet them. Finally, return to the beginning of the line, and again moving from left to right, perform any additions or subtractions as you encounter them.

The two most common mnemonics to remember the order of operations are PEMDAS and *Please Excuse My Dear Aunt Sally.* In either case, P stands for parentheses, E for exponents, M and D for multiplication and division, and A and S for addition and subtraction.

A multiplier in front of parentheses means that everything in the parentheses is to be multiplied by that number. If you can simplify the expression in the parentheses and then multiply, that's great. If not, use the distributive property. Remember that a minus sign in front of the parentheses, as in $13 - (2 + 5)$, acts as -1. If you simplify in the parentheses first, $13 - (2 + 5) = 13 - 7 = 6$, but if you distribute, think of the minus sign as -1.

$$13 - (2 + 5) = 13 - 1(2 + 5) = 13 - 2 - 5 = 11 - 5 = 6$$

Find the value of each expression.

1. $18 - 3^2$

2. $(18 - 3)^2$

3. $15 - 8 + 3$

4. $15 - (8 + 3)$

5. $5^2 - 3 \cdot 2 + 4$

6. $(22 - 7) \cdot 2 + 15 \div 5 - (4 + 8)$

7. $9 - 2 + 4(3 - 5) \div 2$

8. $9 - (2 + 4)(3 - 5) \div 2$

9. $8 + 4^2 - 12 \div 3$

10. $\dfrac{10 + (-8) + [12 - (-4)]}{3(1 - 4)}$

Using variables

Variables are letters or other symbols that take the place of a number that is unknown or may assume different values. You used the idea of a variable long before you learned about algebra. When you put a number into the box in $4 + \square = 6$, or knew what the question mark stood for in $3 - ? = 2$, or even filled in a blank, you were using the concept of a variable. In algebra, variables are usually letters, and determining what number the variable represents is one of your principal jobs.

When you write the product of a variable and a number, you traditionally write the number first, without a times sign, that is, $2x$ rather than $x \cdot 2$. The number is called the coefficient of the variable. A numerical coefficient and a variable (or variables) multiplied together form a term. When you want to add or subtract terms, you can only combine like terms, that is, terms that have the same variable, raised to the same power if powers are involved. When you add or subtract like terms, you add or subtract the coefficients. The variable part doesn't change.

Translating verbal phrases into variable expressions is akin to translating from one language to another. Phrases are translated word by word at first, with care to observe syntax. Variables and numbers are nouns, operation signs $(+, -, \times, \div)$ act as conjunctions, and equal signs or inequality signs are verbs. Sometimes you need to learn an idiom. For example, "3 less than a number" doesn't involve a less-than inequality sign. It translates to $x - 3$.

Simplify each expression by combining like terms where possible.

1. $3t + 8t$

2. $10x - 6x$

3. $5x + 3y - 2x$

4. $2y - 3 + 5x + 8y - 4x$

5. $6 - 3x + x^2 - 7 + 5x - 3x^2$

6. $(5t + 3) + (t - 12r) - 8 + 9r + (7t - 5)$

7. $(5x^2 - 9x + 7) + (2x^2 + 3x + 12)$

8. $(2x - 7) - (y + 2x) - (3 + 5y) + (8x - 9)$

9. $(3x^2 + 5x - 3) - (x^2 + 3x - 4)$

10. $2y - (3 + 5x) + 8y - (4x - 3)$

Write a variable expression for each phrase. Use the variable shown in parentheses at the end of the phrase.

11. Two more than 3 times a number (*x*)

12. Three times a number decreased by 7 (*y*)

13. The quotient of a number and 3, increased by 11 (*t*)

14. Eight less than the product of a number and 9 (*n*)

15. The sum of a number and its opposite (*w*)

16. Three less than 5 times a number, divided by the square of the number (*p*)

17. The square of a number reduced by 4 times the number (*r*)

18. Eight more than the quotient of a number and 1 less than twice the number (*x*)

19. The product of 2 more than 3 times a number and 6 less than 4 times the number (*z*)

20. The square root of 4 times the square of a number decreased by 1 (*v*)

Evaluating expressions

Evaluate, if you take the word back to its roots, means to bring out or lead out the value. You have evaluated, or found the value of, an algebraic expression when you know what numbers the variables stand for. You solve an equation or inequality to find out what the value of the variable is. To evaluate an algebraic expression, replace each variable with its value and simplify according to the order of operations.

To evaluate $\dfrac{5x-3y^2}{4(2x^2-y)}$ when $x=2$ and $y=1$, replace each variable with its value: $\dfrac{5(2)-3(1)^2}{4[2(2)^2-(1)]}$.

The order of operations calls for parentheses first, but don't forget that the fraction bar acts like parentheses, so evaluate the numerator, evaluate the denominator, and then divide.

$$\frac{5(2)-3(1)^2}{4[2(2)^2-(1)]}=\frac{5(2)-3}{4[2(2)^2-(1)]}=\frac{10-3}{4[2(2)^2-(1)]}=\frac{7}{4[2(2)^2-(1)]}$$

$$=\frac{7}{4[2(2)^2-(1)]}=\frac{7}{4[2(4)-(1)]}=\frac{7}{4(8-1)}=\frac{7}{4(7)}=\frac{7}{28}=\frac{1}{4}$$

EXERCISE

1·6

Evaluate each expression for the given values of the variables.

1. $3x-7$ for $x=7$

2. $14-5x$ for $x=6$

3. x^2+2x-7 for $x=-4$

4. $\dfrac{9x+8}{x+11}$ for $x=2$

5. $3x^2-5x+13$ for $x=2$

6. $x+2y$ for $x=7, y=-3$

7. $5x-3y$ for $x=11, y=-10$

8. $(2x^2+5)(4-y)$ for $x=-3, y=-6$

9. $\dfrac{-7x+6}{5-3y}$ for $x=3, y=0$

10. $-4x^2+5xy-3y^2$ for $x=-1, y=2$

Calculator notes #1

A graphing calculator adds many helpful functions to your mathematical toolbox, but at its heart, it's still a calculator and can do all the things you're accustomed to using a calculator to do. Some of them will work exactly the same way, and others will have some differences. Depending on what your previous calculator was like, the difference may be large or small. One of the principal differences is the larger display. It allows you to see several lines of entries and results, and to recall and edit them without a lot of retyping.

To evaluate an expression like $[(22-37)^2 + 15 \div 5 - (4-8)]$ on a traditional calculator, you might do it piece by piece, copying results from the calculator display, retyping a larger portion, and thinking about the order of operations. While that may not technically be necessary, even on a traditional calculator, it's difficult to be sure you've entered complex expressions correctly if they scroll away as you type. The larger display on the graphing calculator allows you to see and edit what you have typed, and the calculator is programmed to execute the order of operations correctly.

Things to be aware of:

- Use all parentheses—not brackets—when you type in the expression.

```
((22-37)²+15/5-(▸
                232
■
```

- An arrowhead at the end of a line indicates that the expression you've typed continues off-screen. Move your cursor up to that line and press $\boxed{ENTER}$, and a copy will appear. You can cursor right and left to see the entire expression and edit it if necessary.
- You can also use the ENTRY feature, which calls back the last thing you typed, by pressing $\boxed{2nd}$ and $\boxed{ENTER}$.
- If your cursor moves up into a little box when you type the exponent, move it forward to get out of the box before typing the rest of the expression.

```
27^3+5*7-42
        1.881676423 E-6
27^3+5*7-42
                19676
```

- Most graphing calculators will allow you to type in a fraction by simply typing numerator/denominator, for example, 3/4. Many now display fractions in "stacked" form: $\frac{3}{4}$. If typing 3/4 and pressing $\boxed{ENTER}$ produces a decimal, check the $\boxed{MODE}$ menu to change settings.

- If you're using a long decimal or complicated fraction, you can store the number and refer to it by a letter. Type the number, press $\boxed{\text{STO}}$, then $\boxed{\text{ALPHA}}$ and the key over which the letter appears. Type $\boxed{\text{ALPHA}}$ and the key for that letter to use the number.
- The letter keys or the $\boxed{x, T, \theta, n}$ keys hold the last value that was stored there. If nothing has been stored, they will be equal to 0, but they do not reset to 0 if something has been stored.

Linear equations

Linear equations take their name from the fact that their graph is a line. They are equations that contain the first power of a variable. No exponents or radicals are involved, and no variables show up in denominators. Linear equations can be simplified to the form $ax + b = 0$. Examples of simplified linear equations include $3x + 7 = 0$ and $\frac{-2}{3}t - \frac{8}{9} = 0$, but before being simplified, they might look like $4x - 9 = 2x + 1$ or $-5(8 - y) = 2(y + 4) - 9$. If an equation has more than one variable term and one constant term on either side of the equal side, take the time to simplify it before starting to solve.

Solving a linear equation is an inverse, or undoing, process. In the equation $3x + 7 = 0$, the variable x was multiplied by 3 and then 7 was added. Solving the equation involves performing opposite operations—subtraction and division—in the opposite order.

Addition and subtraction equations

If $y + 4 = 7$, then you can find the value of y that makes the equation true by subtracting 4 from both sides of the equation to undo the addition.

$$y + 4 - 4 = 7 - 4$$
$$y = 3$$

If the equation was formed by adding, you solve it by subtracting. If it was formed by subtracting, you solve it by adding. The equation $p - 7 = 2$ can be solved by adding 7 to both sides.

$$p - 7 + 7 = 2 + 7$$
$$p = 9$$

EXERCISE

2·1

Solve each equation by adding or subtracting the appropriate number to both sides.

1. $x + 8 = 12$

2. $y - 5 = 11$

3. $t + 3 = 6$

4. $w - 13 = 24$

5. $x + \dfrac{1}{2} = \dfrac{5}{2}$

6. $z - 2.8 = 10.3$

7. $y - 4\tfrac{3}{5} = 7\tfrac{1}{2}$

8. $x + 14 = 8$

9. $y - 7 = -4$

10. $t + 3 = -4$

Multiplication and division equations

The key to solving any equation is getting the variable all alone on one side of the equation. You isolate the variable by performing inverse operations. If the variable has been multiplied by a number, you solve the equation by dividing both sides by that number.

$$-5x = 35$$

$$\dfrac{\cancel{-5}x}{\cancel{-5}} = \dfrac{35}{-5}$$

$$x = -7$$

If the variable has been divided by a number, you multiply both sides by that number to find the value of the variable.

$$\dfrac{x}{6} = -2$$

$$\dfrac{x}{\cancel{6}} \cdot \cancel{6} = -2 \cdot 6$$

$$x = -12$$

EXERCISE 2·2

Solve each equation by multiplying or dividing both sides by the appropriate number.

1. $8x = 32$

2. $\dfrac{z}{7} = 9$

3. $-5y = 42$

4. $\dfrac{t}{9} = -4$

5. $1.5x = 45$

6. $\dfrac{w}{1.1} = 14$

7. $\dfrac{4}{5}t = \dfrac{15}{32}$

8. $\dfrac{m}{4} = -3.1$

9. $-1.3x = 3.9$

10. $\dfrac{z}{5} = 35$

Two-step equations

Most equations require two or more operations to find a solution. The equation $4x - 5 = 19$ says that if you start with a number x, multiply it by 4, and then subtract 5, the result is 19. To solve for x, you will need to perform the opposite, or inverse, operations in the opposite order. You are undoing, stripping away, what was done to x and working your way back to where things started. Undo the subtraction by adding 5.

$$4x - 5 = 19$$
$$4x = 19 + 5$$
$$4x = 24$$

Then undo the multiplication by dividing by 4.

$$4x = 24$$
$$x = \frac{24}{4}$$
$$x = 6$$

EXERCISE 2·3

Solve each equation.

1. $3x - 7 = 32$
2. $-5t + 9 = 24$
3. $4 - 3x = -11$ (Rewrite as $-3x + 4 = -11$ if that's easier.)
4. $9 + 3x = 10$
5. $\dfrac{x}{4} - 7 = -4$

6. $-3x + 5 = -16$
7. $\dfrac{x}{2} + 7 = 15$
8. $11 - 3x = 9.5$
9. $\dfrac{x}{4} - \dfrac{3}{2} = \dfrac{19}{4}$
10. $2x - 7 = -23$

Variables on both sides

When variable terms appear on both sides of the equation, add or subtract to eliminate one of them. This should leave a one- or two-step equation for you to solve.

$$3x_{-2x} - 7 = 2x_{-2x} + 4$$
$$x - 7 = 4$$
$$x = 11$$

EXERCISE 2·4

Eliminate the extra variable term by adding or subtracting; then solve the equation.

1. $5x - 8 = x + 12$
2. $11x + 18 = 3x - 14$
3. $3x + 8 = 4x - 9$

4. $9 - 4x = 16 + 3x$
5. $2x - 5 = 3 - 4x$
6. $8x - 17 = 12 + 3x$

7. $x - 3 = 2 - x$

8. $1.5x - 7.1 = 8.4 + x$

9. $7 - 9x = 7x - 19$

10. $-5x + 21 = 27 - x$

Simplifying before solving

If the equation contains parentheses or has more than two terms on either side, take the time to simplify each side of the equation before you try to solve. If there is a multiplier in front of the parentheses, use the distributive property to multiply and remove the parentheses. Focus on one side at a time and combine like terms. There should be no more than one variable term and one constant term on each side of the equation when you start the process of solving by inverse operations.

$$4(3x - 8) = 10x - (2x + 3) - 1$$
$$12x - 32 = 10x - 2x - 3 - 1$$
$$12x_{-8x} - 32 = 8x_{-8x} - 4$$
$$4x - 32_{+32} = -4_{+32}$$
$$4x = 28$$
$$x = \frac{28}{4} = 7$$

EXERCISE
2·5

Simplify the left side and the right side of each equation. Leave no more than one variable term and one constant term on each side. Then solve each equation.

1. $5(x + 2) = 40$

2. $4(x - 7) + 6 = 18$

3. $5(x - 4) = 7(x - 6)$

4. $4(5x + 3) + x = 6(x + 2)$

5. $8(x - 4) - 16 = 10(x - 7)$

6. $6(2x + 9) - 30 = 4(7x - 2)$

7. $7(x - 1) + 2x = 12 + 5(x + 1)$

8. $6(x - 1) - 2x = 2(x + 1) + 4(2 - x)$

9. $5(6x + 2) + 7(4 - 12x) = 35 - (6 + 27x)$

10. $8(2x - 5) - 2(x - 2) = 5(x + 7) - 4(x + 8)$

Absolute value equations

An absolute value equation, with a variable expression inside the absolute value signs, isn't actually a linear equation, but it's closely related. The graphs of absolute value functions aren't lines; they're V-shaped. Absolute value equations will usually have two solutions. The expression between the absolute value signs may be equal to the expression on the other side of the equal sign, or it may be equal to the opposite of that expression.

$$|2x - 7| = 5$$

$$\begin{array}{ccc} 2x - 7 = 5 & & 2x - 7 = -5 \\ 2x = 12 & \text{or} & 2x = 2 \\ x = 6 & & x = 1 \end{array}$$

Always isolate the absolute value before considering the two cases. All terms other than the absolute value should be moved to the other side before you say that the expression in the absolute value signs could equal the other side or its opposite. If there are two or more terms on that side, be sure to form the opposite correctly by distributing the negative.

$$4|3x-1|+3=5x+13$$
$$4|3x-1|=5x+10$$
$$|3x-1|=\frac{5x+10}{4}$$

$$3x-1=\frac{5x+10}{4} \qquad \text{or} \qquad 3x-1=-\frac{5x+10}{4}$$
$$12x-4=5x+10 \qquad\qquad\qquad 12x-4=-5x-10$$
$$7x-4=10 \qquad\qquad\qquad\quad 17x-4=-10$$
$$7x=14 \qquad\qquad\qquad\qquad 17x=-6$$
$$x=2 \qquad\qquad\qquad\qquad\quad x=\frac{-6}{17}$$

EXERCISE
2·6

Each equation involves absolute value. Solve each equation by the method above.

1. $|3x + 5| = 23$

2. $|6x - 3| = 17$

3. $|5x + 2| = 47$

4. $|3 + 6x| = 33$

5. $|7 + 8x| = 51$

6. $|30 + 3x| = 18x$

7. $|40 - 2x| = 6x$

8. $|3x - 11| = 8 + x$

9. $|8x - 2| = 2x + 22$

10. $|9x + 2| - 3x = 17 + x$

Mixture problems: coffee, coins, and cars

Many of the problems you encounter in algebra fall into common types, and if you recognize the type and have a strategy for dealing with it, the problems are easier to solve. Often these problems are about mixing things. A merchant wants to mix different types of coffees or nuts. An amount of money is a mixture of different denominations of coins or bills. A laboratory is mixing chemicals in a solution, or a theater is offering different admission prices for adults and children. Even the famous "Two trains leave Chicago …" problems are mixing the distances traveled by the trains (or cars).

When you're faced with a problem from this mixture family, you may find it easier to solve if you organize the information in a chart before you try to write and solve an equation. If a merchant is blending 10 lb of coffee, for example, you want a row for each of the individual coffees and the blend, and columns for pounds of coffee, price per pound, and total value. If you call the amount of one coffee x and the other $10 - x$ and you enter the per-pound price of each, you can multiply across to get the total value. The total value of the individual coffees should add up to the total value for the mixture, so your equation comes from adding down the last column.

	Pounds of coffee	Price per pound	Total value
Colombian	x	$4.95	$4.95x$
Arabica	$10 - x$	$2.80	$2.8(10 - x)$
Blend	10	$3.75	$37.50

For coin problems, the column headings are number of coins and value of the coins, and the equation is formed the same way.

	Number of coins	Value of coins	Total value
Dimes	x	$0.10	$0.1x$
Quarters	$12 - x$	$0.25	$0.25(12 - x)$
Blend	12		$1.95

$$0.1x + 0.25(12 - x) = 1.95$$
$$0.1x + 3 - 0.25x = 1.95$$
$$-0.15x + 3 = 1.95$$
$$-0.15x = -1.05$$
$$x = 7$$

You end up with 7 dimes and 5 quarters.

The cars, trains, and planes take a little more analysis, but the setup of the problem is very much the same. You want a row in the table for each vehicle, and your column headings come from a familiar formula: rate (of speed) times time equals distance. Suppose two trains leave Chicago, traveling in opposite directions. One travels at 80 mph and the other at 75 mph. When will the trains be 1085 mi apart? Let x be the time it takes for this to happen.

	Rate	Time	Distance
Train 1	80	x	$80x$
Train 2	75	x	$5x$
			1085

Here's where the little bit of extra thinking comes in. What do you do with the distances? Since the trains are going in opposite directions, you add the distances. If they were traveling in the same direction at different speeds and you wanted to know how far ahead the faster one had gotten, you'd subtract the distances they'd traveled. Draw a picture to help you imagine what's happening in a particular problem.

EXERCISE
2·7

Solve each problem by writing and solving an equation.

1. Jake had 12 coins in his pocket, totaling 95 cents. If the coins were all dimes and nickels, how many nickels did Jake have?

2. Two cars leave Omaha at the same time. One travels east at 55 mph and the other travels west at 65 mph. When are the cars 500 mi apart?

3. You decide to make 10 lb of a peanut-and-raisin mixture to sell at the class snack sale. You can buy peanuts for $2.50 per pound and raisins for $1.75 per pound. If you want to sell the mixture for $2 per pound, how many pounds of peanuts and how many pounds of raisins should you use?

4. The Fletcher family puts loose change into a jar every evening. Once a month, the family counts the coin and prepares to take it to the bank, but removes all the pennies first. Last month, they found there were 14 more quarters than dimes, and the number of nickels was 8 less than 4 times the number of dimes. If the total value of the nickels, dimes, and quarters was $15.75, how many dimes did they have?

5. Jessie and her friends pack a tailgate picnic and, at exactly 1 p.m., set out for the football game, driving at 35 mph. Half an hour after they leave the house, Jessie's mom notices their picnic basket, fully packed, sitting on the driveway. She grabs the basket, jumps in her car, and drives at 40 mph, the legal limit. When will Mom catch up with Jessie?

6. Bert and Harriet collect quarters and pennies. When they wrapped coins to take to the bank, they had $42.50. If they wrapped a total of 410 coins, how many were pennies?

7. Admission to the school fair is $2.50 for students and $3.75 for others. If 2848 admissions were collected for a total of $10,078.75, how many students attended the fair?

8. Alisha designs a new tea blend by mixing Sweet Rose Tulsi tea with Orange Blossom green tea. Sweet Rose Tulsi tea sells for $4.98 an ounce and Orange Blossom green tea sells for $1.98 an ounce. If she wants to blend 16 ounces of the tea blend to sell for $4.23 per ounce, how many ounces of each tea should she use?

9. At precisely noon, one plane leaves New York, heading for Orlando, and another leaves Orlando, heading for New York. The distance from New York to Orlando is 1300 mi. The plane from New York flies at 450 mph and the Orlando plane flies at 490 mph. When will the planes be 125 mi apart?

10. I leave my family's vacation cabin at 8 a.m. and start driving home at a nice, safe 45 mph. Two hours later, my husband, who always drives as fast as the law will allow, leaves the cabin and starts driving home at 65 mph. When can I expect him to pass me?

Calculator notes #2

To check the solution of an equation (or to approximate the solution(s) of an equation beyond your current skills), you can use the graphing functions of your calculator.

- Start by moving all the terms of the equation to one side, equal to zero. Simplify if you wish, but the calculator won't care if you leave it messy.
- Graph the non-zero side by pressing $\boxed{Y=}$, clearing any previous equations, and typing in the non-zero side of your equation.

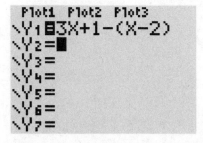

- Press ⌊ZOOM⌋ and choose 6: ZStandard to set to the standard viewing window.

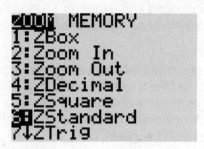

- Look to see where the graph crosses the *x*-axis. The *x*-intercept is where the expression you typed equals zero. (For certain types of equations, there may be more than one.) If the equation has no solution, there will not be an *x*-intercept, but before you decide that, zoom out or change window.
- Once you've located the *x*-intercept, move your cursor to the left of it. Press ⌊2nd⌋ ⌊TRACE⌋, and choose 2: ZERO. Make sure your cursor is left of the *x*-intercept, press ⌊ENTER⌋.

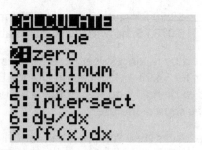

- Move to the right of the *x*-intercept, press ⌊ENTER⌋.
- Move close to the *x*-intercept, press ⌊ENTER⌋.

The coordinates of the *x*-intercept will display at the bottom of the screen. The *x*-coordinate is the solution of the equation.

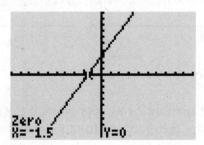

Linear inequalities

3

From the beginning of your training in arithmetic, you used an equal sign (=) to signify several different things: "when you perform this operation you get this result" or "these two expressions have the same value" and others. You may well have taken for granted the equal sign could be read both ways. You saw $5 + 3 = 8$ as identical to $8 = 5 + 3$. Mathematicians say the relation of equality is symmetric, and not every relationship is. When you meet inequalities, the relationships of "is greater than" or "is less than" in particular, you have relationships that are not symmetric. For example, $7 < 12$ is certainly not the same as $12 < 7$. One is true; one is not. These not-symmetric relationships bring some new rules.

When you solve an equation, you find the value of the variable that makes the two sides of the equation identical. We think of equations as having one solution, although there are exceptions to that rule. Inequalities, or sentences that involve a variable and say that two expressions are not equal, ask you to find the values that make one side larger than the other. The solution will be a set of numbers, rather than the single value that solves a linear equation. If you think of even the simplest inequality, like $x > 4$, there are infinitely many numbers that could replace x and make a true statement.

Simple inequalities

Linear inequalities can be solved in much the same way as linear equations with one important exception. When you multiply or divide both sides of an inequality by a negative number, the direction of the inequality sign reverses. Imagine a number line for a moment, and suppose you don't know the value of x but you do know that the opposite of x is greater than 5. It could be 6, or 10, or 152, or any number to the right of 5 on the number line. If the opposite of x, or $-x$, were 6, x itself would be -6. If $-x$ were 10, x would be -10. If $-x$ were 152, x would be -152, and so on. Now think about where those values of x lie on the number line. If $-x$ is any number to the right of 5, x is a number to the left of -5. If $-x > 5$, then $x < -5$. Remember that the positive and negative sides of the number line are mirror images of one another. When you multiply or divide both sides of an inequality by a negative number, you go through the looking glass and the larger-smaller relationship flips.

Follow the same plan for solving a simple inequality as you would for a simple equation. Simplify the left side, then the right side, until you have no more than one variable term and one constant term on a side. Eliminate one variable term and one constant term by adding or subtracting on both sides. Remember you want the variable on one side and the constant on the other.

It's only at the last step that you have to think about it a little differently. Your last step is to divide both sides by the coefficient of the variable. If that coefficient is a negative number, you'll also reverse the direction of the inequality sign: > will become <, and < will become >. If there's an "or equal to," it will stay there but the direction of the < or > will change. Here's an example.

Simplify the left side: $-5x - 2 + 3(x - 1) < x + 2(3x - 1)$
$$-5x - 2 + 3x - 3 < x + 2(3x - 1)$$
$$-2x - 5 < x + 2(3x - 1)$$
Simplify the right side: $\qquad -2x - 5 < x + 6x - 2$
$$-2x - 5 < 7x - 2$$
Eliminate a variable term: $-9x - 5 < -2$
Eliminate a constant term: $-9x < 3$
Divide both sides by -9 and change the < to >: $x > \dfrac{3}{-9}$
The solution is $x > -\dfrac{1}{3}$

The solution set of an inequality can be graphed on the number line by shading the appropriate portion of the line as shown in Figure 3.1. Use an open circle to mark the boundary of the solution set if the inequality sign is < or >, and a solid dot for inequalities containing ≤ or ≥.

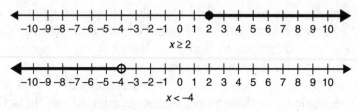

Figure 3.1 Solutions of inequalities graphed on the number line.

EXERCISE
3·1

Solve each inequality and graph the solution set on a number line.

1. $3x - 5 \geq 22$

2. $2x - 5 > 13 - 4x$

3. $3x + 2 \leq 8x + 22$

4. $12x + 3 < x + 36$

5. $t - 9 \geq 24 - 10t$

6. $2y - 13 > 4(2 - y)$

7. $5x + 10(x - 1) \geq 95$

8. $5x - 4 \leq 13x + 28$

9. $3x - 2 < 2x - 3$

10. $-x + 5 \geq -2 + x$

Compound inequalities

A compound inequality can be written as two linear inequalities joined by the conjunction *and* or the conjunction *or*. The sentence "$3x - 7 \leq 4x + 2$ or $x - 3 > 9$" is an example of a compound inequality using "or." The sentence "$2(3x + 5) \geq 3x - 19$ and $x + 11 \leq 2x - 9$" is a compound inequality using "and." Some *and* inequalities can be written in a condensed form; for example, the compound inequality $7 > 5x + 2$ and $5x + 2 > -13$ can be written as $7 > 5x + 2 > -13$. One key to using this condensed form is the repetition of the same expression (in this case, $5x + 2$) in each inequality as a link between them. You also need to have the inequality signs pointing in the same direction. The condensed form $7 > 5x + 2 > -13$ works, and $-13 < 5x + 2 < 7$ would as well, but $7 > 5x + 2 < -13$ does not. To solve compound inequalities, rewrite them as two inequalities, solve each inequality separately, and join the two solutions with the same conjunction.

$$22 > 7 - 5x \geq -3$$

$22 > 7 - 5x$	and	$7 - 5x \geq -3$
$15 > -5x$	and	$-5x \geq -10$
$-3 < x$	and	$x \leq 2$

$$-3 < x \leq 2$$

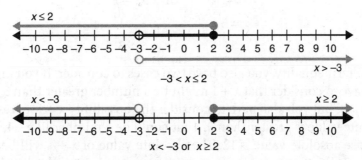

Figure 3.2   Graphing the solution set of a compound inequality.

To graph the solution set of a compound inequality, graph the solution sets of the two component inequalities, shading lightly. If the inequalities are connected with the conjunction *and*, the solution set is the intersection or overlap of the two graphs. You want the numbers that make both the first inequality AND the second inequality true. For compound inequalities that are connected by *or*, you want to keep both areas of shading. The solution set includes the numbers that solve one inequality OR the other OR both. If the two shaded areas for an *or* inequality overlap, you may find that you have only one shaded area (see Figure 3.2).

EXERCISE 3·2

Solve each inequality and graph the solution set on a number line.

1. $x - 10 > 30$ or $x - 4 < -10$

2. $6 < y - 9 < 15$

3. $-3 \leq 4x + 5 < 2$

4. $6x - 4 \geq 26$ or $3x + 8 < 14$

5. $-43 < 11x + 1 \leq 12$

6. $8 + 3x > 4 - 3x$ or $9x + 12 < -87 - 2x$

7. $-10 \leq 15y + 5 \leq 6$

8. $-13 < 47 - 3x < -2$

9. $3x + 4 < 6x + 7$ or $5x - 2 > 3x + 18$

10. $-4y + 31 < y + 16 \leq 3y + 2$

Absolute value inequalities

When you first learn about the idea of absolute value, you describe it in different ways: the number without its sign or a distance without regard to direction. You learn that $|7| = 7$ but $|-7| = 7$ as well. Every number has an absolute value, but no absolute value is ever negative. Every absolute value (except 0) may trace back to either a positive or a negative number. That's important when you start to introduce variables. It's easy to see the value of x if you're thinking about $|-12| = x$, but if you're trying to determine x in $|x| = 5$, you need to realize that x could be 5 or −5.

In an absolute value equation, you look for two solutions, unless the absolute value is equal to zero. You consider the possibility that the expression within the absolute value signs is equal to the number on the other side of the equation and the possibility that the expression is equal to the opposite of that number. If you need to solve $|3x - 8| = 7$, you actually have to solve two equations.

$$|3x - 8| = 7$$
$$3x - 8 = 7 \qquad 3x - 8 = -7$$
$$3x = 15 \quad \text{or} \quad 3x = 1$$
$$x = 5$$
$$x = \frac{1}{3}$$

In an absolute value inequality, you also have two cases to consider. If you know, for example, that $|x + 1| > 3$, you have to consider that $x + 1$ might be a number greater than 3, like 4 or 12 or 107, so you'd solve $x + 1 > 3$. But you also have to consider the possibility $x + 1$ represents a negative number whose absolute value is greater than 3. That could be a number like −4, whose absolute value is 4, or −12, whose absolute value is 12. The absolute value of $x + 1$ will be greater than 3 when $x + 1$ is less than −3. The two possibilities you have to examine to solve $|x + 1| > 3$ are $x + 1 > 3$ and $x + 1 < -3$. But only one of those can be true at any time, so $|x + 1| > 3$ translates to the compound inequality $x + 1 < -3$ or $x + 1 > 3$. If your inequality says $|x - 2| < 7$, $x - 2$ might be a positive number less than 7, like 3, or a negative number greater than − 7, like −1 (or zero). But if you try some other numbers, like −10, because it's less than 7, or 12, because it's greater than −7, you find those don't work. Only numbers that are both less than 7 and greater than −7 will work. The inequality $|x - 2| < 7$ is equivalent to $- 7 < x - 2 < 7$.

To solve an inequality that involves an absolute value, first isolate the absolute value. Simplify the inequality until one side is only an absolute value expression, and the other is as simple as possible. Ideally, it will be just a constant, but if there is a variable term as well, simplify as much as you can.

Rewrite the absolute value inequality as a compound inequality. If the absolute value is greater than the expression on the other side, it will become an *or* inequality. Absolute values less than an expression translate to an *and* inequality. The mnemonic "GreatOR, Less ThAND" will help you remember. Translate using the format:

Opposite of constant	Inequality sign	Expression within the absolute value	Conjunction (AND or OR)	Expression within the absolute value	Inequality sign	Constant

Here are a couple of examples.

$$50 - 7|3x+2| < 15$$
$$-7|3x+2| < 15 - 50$$
$$-7|3x+2| < -35$$
$$|3x+2| > \frac{-35}{-7}$$
$$|3x+2| > 5$$

$$4|3-7x| + 8 \leq 48$$
$$4|3-7x| \leq 48 - 8$$
$$4|3-7x| \leq 40$$
$$|3-7x| \leq \frac{40}{4}$$
$$|3-7x| \leq 10$$

$$|3x+2| > 5 \quad \text{becomes} \quad -5 > 3x+2 \text{ or } 3x+2 > 5$$

$$-5 > 3x+2 \text{ OR } 3x+2 > 5$$

$$-5 - 2 > 3x \qquad \qquad 3x > 5-2$$
$$-7 > 3x \qquad \text{OR} \qquad 3x > 3$$
$$\frac{-7}{3} > x \qquad \qquad x > \frac{3}{3}$$
$$\qquad \qquad x > 1$$

$|3 + 7x| \leq 10$ becomes $-10 \leq 3 - 7x$ AND $3 - 7x \leq 10$, or, if you prefer, $-10 \leq 3 - 7x \leq 10$.

$$-10 \leq 3 - 7x \text{ AND } 3 - 7x \leq 10$$

$$-10 - 3 \leq -7x \qquad \qquad$$
$$-13 \leq -7x \qquad \qquad -7x \leq 10 - 3$$
$$\qquad \qquad -7x \leq 7$$
$$\frac{-13}{-7} \geq x \qquad \text{AND} \qquad x \geq \frac{7}{-7}$$
$$\frac{13}{7} \geq x \qquad \qquad x \geq -1$$

$$\frac{13}{7} \geq x \geq -1$$

3·3

Solve each inequality and graph the solution set on a number line.

1. $|2x-7| < 9$

2. $|3x+5| \geq 17$

3. $|3x+5| + 4 > 36$

4. $|4-5x| - 3 \leq 2$

5. $2|7x-12| - 5 > 13$

6. $-4|9-x| + 7 < -17$

7. $2 - |11x+5| \leq -36$

8. $4|17-2x| - 9 \geq 19$

9. $|6x-11| < -x+3$

10. $|13-5x| \geq 2x-1$

Coordinate graphing

The graph of an equation in two variables gives a picture of all the pairs of numbers that balance the equation. Studying the graph will help you understand the relationship between the variables and can sometimes help you find the solution of an equation.

The coordinate plane

The Cartesian coordinate system, named for René Descartes, is a rectangular coordinate system that locates every point in the plane by an ordered pair of numbers (x, y). The x-coordinate indicates horizontal movement, and the y-coordinate vertical movement. Movement begins from a point $(0, 0)$, called the origin, where two number lines, one horizontal and one vertical, intersect. The horizontal number line is the x-axis and the vertical is the y-axis. Positive x-coordinates are to the right of the origin, and negative x-coordinates to the left. A y-coordinate that is positive is above the x-axis, and a negative y-coordinate is below (see Figure 4.1).

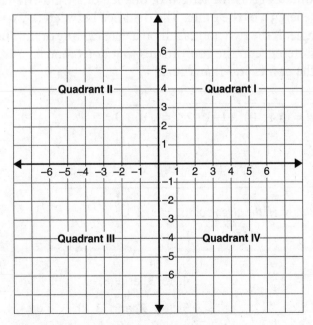

Figure 4.1 The coordinate plane divided into four quadrants.

The *x*- and *y*-axes divide the plane into four quadrants. The first quadrant is the section in which both the *x*- and *y*-coordinates are positive, and the numbering of the quadrants goes counterclockwise.

Plot each point on the coordinate plane.

1. A(−4, 2)

2. B(8, −3)

3. C(0, 5)

4. D(−2, −6)

5. E(−5, 0)

Tell which quadrant contains the point.

6. (−4, 5)

7. (3, 2)

8. (4, −1)

9. (−2, −2)

10. (3, −4)

Distance

The distance between two points can be calculated by means of the distance formula $d = \sqrt{(x_2 - x_1)^2 + (y_2 - y_1)^2}$. The formula is an application of the Pythagorean theorem, in which the difference of the *x*-coordinates gives the length of one leg of a right triangle, and the difference of the *y*-coordinates the length of the other. The distance between (x_1, y_1) and (x_2, y_2) is the hypotenuse of the right triangle. If the two points fall on a vertical line or on a horizontal line, the distance will simply be the difference in the coordinates that don't match.

The distance between the points (4, −1) and (0, 2) is

$$d = \sqrt{(4-0)^2 + (-1-2)^2} = \sqrt{16+9} = 5$$

Find the distance between the given points.

1. (4, 5) and (7, −4)

2. (6, 2) and (7, 6)

3. (−7, −1) and (−5, −6)

4. (5, 3) and (8, −2)

5. (−4, 2) and (3, 2)

Given the distance between the two points, find the possible values for the missing coordinate.

6. $(a, -2)$ and $(7, 2)$ are 5 units apart.

7. $(-1, 3)$ and $(4, d)$ are 13 units apart.

8. $(8, -6)$ and $(c, -6)$ are 7 units apart.

9. $(2, b)$ and $(2, -1)$ are 9 units apart.

10. (a, a) and $(0, 0)$ are $4\sqrt{2}$ units apart.

Midpoints

The midpoint of the segment that connects (x_1, y_1) and (x_2, y_2) can be found by averaging the x-coordinates and averaging the y-coordinates.

$$M = \left(\frac{x_1 + x_2}{2}, \frac{y_1 + y_2}{2} \right)$$

The midpoint of the segment connecting $(4, -1)$ and $(0, 2)$ is

$$M = \left(\frac{4+0}{2}, \frac{-1+2}{2} \right) = \left(2, \frac{1}{2} \right)$$

EXERCISE
4·3

Find the midpoint of the segment with the given end points.

1. $(2, 3)$ and $(5, 8)$

2. $(-3, 1)$ and $(-1, 8)$

3. $(-5, -3)$ and $(-1, -1)$

4. $(8, 0)$ and $(0, 8)$

5. $(0, -2)$ and $(4, -4)$

Given the midpoint M of the segment connecting A and B, find the missing coordinate.

6. $A(x, 6)$, $B(6, 8)$, $M(4, 7)$

7. $A(-1, 3)$, $B(x, 9)$, $M(3, 6)$

8. $A(-5, y)$, $B(7, -3)$, $M(1, 3)$

9. $A(4, -9)$, $B(-2, y)$, $M(1, -7)$

10. $A(0, 4)$, $B(x, 0)$, $M(8, 2)$

Slope and rate of change

The slope of a line is a measurement of the rate at which it rises or falls. A rising line has a positive slope whereas a falling line has a negative slope as shown in Figure 4.2. The larger the absolute value of the slope, the steeper the line. A horizontal line has a slope of 0, and a vertical line has an undefined slope.

$$m = \frac{\text{rise}}{\text{run}} = \frac{y_2 - y_1}{x_2 - x_1}$$

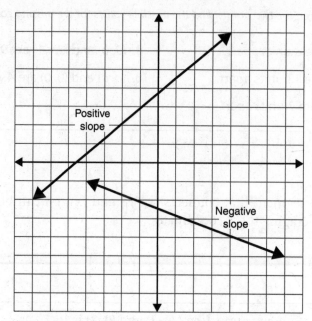

Positive slope

Negative slope

Figure 4.2 Lines with positive and negative slopes.

The slope of the line through the points (4, –1) and (0, 2) is

$$m = \frac{2-(-1)}{0-4} = -\frac{3}{4}$$

EXERCISE

4·4

Find the slope of the line that passes through the two points given.

1. (–5, 5) and (5, –1)   4. (4, 6) and (8, 7)

2. (6, –4) and (9, –6)   5. (7, 2) and (7, 5)

3. (3, 4) and (8, 4)

If a line has the given slope and passes through the given points, find the missing coordinate.

6. $m = -4$, (4, y) and (3, 2)   9. $m = -\frac{3}{5}$, (x, 1) and (2, –5)

7. $m = 2$, (2, 9) and (x, 13)   10. $m = 0$, (–4, y) and (–6, 3)

8. $m = \frac{1}{2}$, (–4, 1) and (3, y)

Graphing linear equations

A linear equation in two variables has infinitely many solutions, each of which is an ordered pair (x, y). The graph of the linear equation is a picture of all the possible solutions.

Table of values

The most straightforward way to graph an equation is to choose several values for x, substitute each value into the equation, and calculate the corresponding values for y. This information can be organized into a table of values. Geometry tells us that two points determine a line, but when building a table of values, it is wise to include several more so that any errors in arithmetic will stand out as deviations from the pattern.

When you build a table of values, make a habit of choosing both positive and negative values for x. Of course, you can chose $x = 0$, too. Usually, you'll want to keep the x-values near 0 so that the numbers you're working with don't get too large. If they do, you'll need to extend your axes, or relabel your scales by 2s or 5s or whatever multiple is convenient. If the coefficient of x is a fraction, choose x-values that are divisible by the denominator of the fraction. This will minimize the number of fractional coordinates, which are hard to estimate.

EXERCISE
4·5

Construct a table of values and graph each equation.

1. $y = 3x + 2$

2. $2y = 4x - 8$

3. $3y = 4x + 12$

4. $x + y = 10$

5. $y - 2x = 7$

6. $6x + 2y = 12$

7. $3x - 4y = 12$

8. $y = \frac{1}{2}x + 5$

9. $y = -\frac{2}{3}x - 2$

10. $4x - 3y = 3$

Slope and *y*-intercept

To draw the graph of a linear equation quickly, put the equation in slope-intercept, or $y = mx + b$, form. The value of b is the y-intercept of the line, and the value of m is the slope of the line. Begin by plotting the y-intercept; then count the rise and run and plot another point. Repeat a few times and connect the points to form a line.

Intercept-intercept

If the linear equation is in standard, or $ax + by = c$, form, it is very easy to find the x- and y-intercepts of the line. The x-intercept is the point at which y equals 0, and the y-intercept is the point at which x equals 0. Substituting 0 for y reduces the equation to $ax = c$, and dividing by a gives the x-intercept. In the same way, substituting 0 for x gives $by = c$, and the y-intercept can be found by dividing by b. Plotting the x- and y-intercepts and connecting them will produce a quick graph.

Use quick graphing techniques to draw the graph of each of the following equations.

1. $y = -\dfrac{3}{4}x + 1$

2. $2x - 3y = 9$

3. $y = -4x + 6$

4. $6x + 2y = 18$

5. $x - 2y = 8$

6. $y = -3x - 4$

7. $y - 6 = 3x + 1$

8. $3x + 5y = 15$

9. $2y = 5x - 6$

10. $3x - 2y - 6 = 0$

Vertical and horizontal lines

Horizontal lines fit the $y = mx + b$ pattern, but since they have a slope of 0, they become $y = b$. Whatever value you choose for x, the y-coordinate will be b.

Vertical lines have undefined slopes, so they cannot fit the $y = mx + b$ pattern, but since every point on a vertical line has the same x-coordinate, they can be represented by an equation of the form $x = c$, where c is a constant. The value of c is the x-intercept of the line.

Identify each line as horizontal, vertical, or oblique.

1. $x = -3$

2. $y = 4$

3. $2x + 8 = 0$

4. $y - 4x = 0$

5. $y + 1 = 4$

Graph each equation.

6. $y = -3$

7. $x = 2$

8. $y = 5$

9. $x = -1$

10. $5y - 18 = 2$

Graphing linear inequalities

Linear inequalities can also be graphed on the coordinate plane. Begin by graphing the line that would result if the inequality sign were replaced with an equal sign. If the inequality is ≥ or ≤, use a solid line. For > or <, use a dotted line. Test a point on one side of the line in the inequality; the origin is often a convenient choice. If the result is true, shade that side of the line; if not, shade the other side (see Figure 4.3).

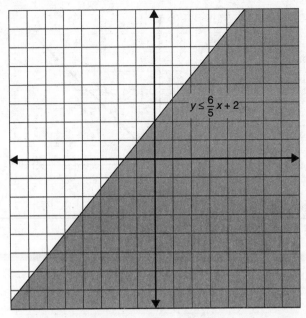

$$y \le \frac{6}{5}x + 2$$

Figure 4.3 Graph of a linear inequality in two variables.

Graph each inequality and indicate the solution set by shading.

1. $y \le 2x - 5$

2. $y > 5x - 4$

3. $y < x$

4. $y \ge \frac{1}{2}x + 1$

5. $y \le -2x + 5$

6. $3x - y > 0$

7. $8y - 3x < -4$

8. $x - y \ge -5$

9. $2x - y \le -2$

10. $2x - 3y \ge -15$

Graphing absolute value equations

Many equations involving absolute value have graphs that are composed of two linear segments that have opposite slopes. The graph of $y = |x|$, for example, is made up of the graph of $y = -x$ when $x < 0$ and the graph of $y = x$ when $x \ge 0$. The result is a V-shaped graph as shown in Figure 4.4.

When you start to build a table of values to graph an absolute value equation, first find the value of x that will make the expression in the absolute value signs equal 0. Choose a few values below it and a few above it to fill out your table.

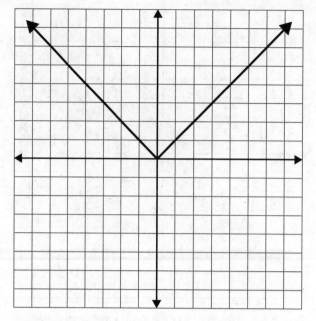

Figure 4.4 Graph of the absolute value equation.

Construct a table of values and graph the equation.

1. $y = |x-4|$

2. $y = |x|-3$

3. $y = 2|x|+1$

4. $y = 2|x+1|$

5. $y = -3|x|$

6. $y = \frac{1}{2}|x+5|$

7. $y = -\frac{1}{3}|x-5|$

8. $y = |x-2|+1$

9. $y = |x+5|-6$

10. $y = -2|x+3|-4$

Writing linear equations

It is sometimes necessary to determine the equation that describes a graph either by looking at the graph itself or by using information about the graph.

Slope and *y*-intercept

If the slope and *y*-intercept of the line are known or can be read from the graph, the equation can be determined easily by using the $y = mx + b$ form. Replace *m* with the slope and *b* with the *y*-intercept.

Point and slope

If the slope is known and a point on the line other than the *y*-intercept is known, the equation can be found by using point-slope form: $y - y_1 = m(x - x_1)$. Replace *m* with the slope, and replace x_1 and y_1 with the coordinates of the known point. Distribute and simplify to put the equation in $y = mx + b$ form.

Two points

If two points on the line are known, the slope can be calculated using the slope formula. Once the slope is found, you can use the point-slope form and fill in the slope and either one of the two points.

EXERCISE

4·10

Write the equation of the line described.

1. Slope = 3 and *y*-intercept (0, 8)

2. Slope = −5 and *y*-intercept (0, 2)

3. Slope = $\dfrac{2}{3}$ and *y*-intercept (0, 6)

4. Slope = 4 and passing through the point (3, 7)

5. Slope = $\dfrac{1}{2}$ and passing through the point (4, 3)

6. Slope = $-\dfrac{3}{2}$ and passing through the point (−4, −1)

7. Passing through the points (0, 3) and (2, 7)

8. Passing through the points (3, 4) and (9, 8)

9. Passing through the points (0, −3) and (3, 1)

10. Passing through the points (2, 3) and (8, −6)

Parallel and perpendicular lines

Parallel lines have the same slope. Perpendicular lines have slopes that multiply to −1, that is, slopes that are negative reciprocals. To find the equation of a line parallel to or perpendicular to a given line, first determine the slope of the given line. Be sure the equation is in slope-intercept form before trying to determine the slope. Use the same slope for a parallel line or the negative reciprocal for a perpendicular line, along with the given point, in point-slope form.

To find a line parallel to $y = 3x - 7$ that passes through the point (4, −1), use the slope of 3 from $y = 3x - 7$ and the point (4, −1) in point-slope form and simplify.

$$y - (-1) = 3(x - 4)$$
$$y + 1 = 3x - 12$$
$$y = 3x - 13$$

To find a line perpendicular to $y = 3x - 7$ that passes through the point (4, −1), use a slope of $-\dfrac{1}{3}$.

$$y - (-1) = -\frac{1}{3}(x - 4)$$
$$y + 1 = -\frac{1}{3}x + \frac{4}{3}$$
$$y = -\frac{1}{3}x + \frac{1}{3}$$

Determine whether the lines are parallel, perpendicular, or neither.

1. $y = \frac{1}{3}x - 2$ and $3x + y = 7$

2. $x - 5y = 3$ and $2x - 10y = 9$

3. $2y - 8x = 9$ and $4y = 3 - 18x$

4. $4y = 6x - 7$ and $y = \frac{3}{2}x + 5$

5. $y = \frac{4}{5}x + 8$ and $5x + 4y = 8$

Find the equation of the line described.

6. Parallel to $y = 5x - 3$ and passing through the point (3, –1)

7. Perpendicular to $6y - 8x = 15$ and passing through the point (–4, 5)

8. Parallel to $4x + 3y = 21$ and passing through the point (1, 1)

9. Perpendicular to $y = 4x - 3$ and passing through the point (4, 13)

10. Parallel to $2y = 4x + 16$ and passing through the point (8, 0)

Calculator notes #3

The primary task of the graphing calculator, of course, is to graph. You practice this with equations that are actually very simple to graph by hand so that you'll have the skills to produce graphs of more complicated equations. The basics are simple. Press $\boxed{Y=}$, clear any old equations, type in your equation, press $\boxed{GRAPH}$. Of course, there can be other things to think about.

Your equation must be in "$y=$" form. If you're trying to graph $3x - 5y = -41$, it's your job to first isolate y.

$$3x - 5y = -41$$
$$-5y = -3x - 41$$
$$y = \frac{-3x - 41}{-5}$$

You don't have to simplify any more than that, and although it probably runs counter to your training in good practice, it might be better not to do any more. As the equation stands now, you can enter it as Y1 = (–3x – 41)/–5. If you simplify, you'll have more fractions, and the possibility of sign errors. Notice that the entire numerator is enclosed in parentheses; don't skip them, because Y1 = –3x – 41/–5 only divides –41 by –5 and produces a dramatically different graph.

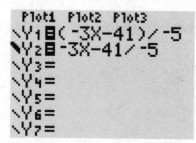
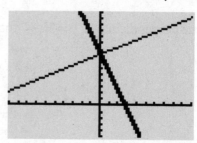

When you've typed in your equation, you can hit GRAPH , but you might want to press ZOOM and choose 6:ZStandard instead. That sets your viewing window to −10 to 10 on the x-axis and −10 to 10 on the y-axis. Hitting GRAPH displays whatever window you used last. That might be fine, but it might not. The window is just a portion of the infinite coordinate plane. If you graph the line $y = x$ but your window is set to look at the portion of the plane for which x is between −300 and −200 and y is between 500 and 600, you're not going to see the line.

If you don't see the graph, or don't see enough of it, first press TRACE and notice the coordinates that display at the bottom of the screen. Those are the coordinates of a point on the line. You need to reset the window so that you can see that point. Also think about the y-intercept. Choose the window that seems right to you. Press WINDOW , set Xmin, Xmax, Ymin, and Ymax. You only need to adjust Xscl and Yscl if you're going to look at a very large window. Set Xscl or Yscl to larger numbers to prevent the tick marks on the axes from blurring together.

Most graphing calculators are not able to graph a vertical line, but may be able to draw one. Press 2nd PRGM to get the DRAW menu, choose 4: Vertical, and enter the x-value. You cannot trace along this line or find its x-intercept because it is not a graph.

Systems of linear equations and inequalities

5

A single equation that contains two or more variables has infinitely many ordered pairs in its solution set. A system of equations is a group of equations, each of which involves two or more variables. For systems with two variables and two equations, the solution is an ordered pair of values that make both of the equations in the system true.

Graphing systems of equations

A system of two linear equations in two variables can be solved by graphing each of the equations and locating the point of intersection. The coordinates of the point of intersection are the values of x and y that solve the system. If the lines are parallel, the system has no solution.

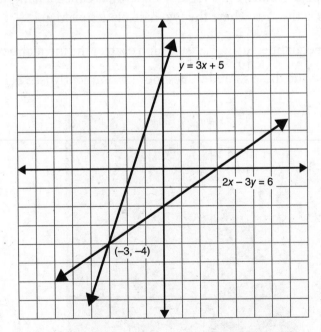

Figure 5.1   Solving a system of equations by graphing.

In Figure 5.1, by graphing $2x - 3y = 6$ and $y = 3x + 5$ on the same set of axes, you can see that they cross at the point $(-3, -4)$. That means that $x = -3$ and $y = -4$ will solve both equations. Each equation has an infinite number of solutions, but this one is the only one they have in common.

Graph each system of equations and locate the solution, if one exists.

1. $x + y = 6$
   $y = 2x$

2. $x - y = -6$
   $2x + y = 0$

3. $x + y = 8$
   $y = 2x - 4$

4. $x + y = -6$
   $x - 2y = 3$

5. $3x + 4y = 12$
   $x + y = 2$

6. $2x + y = 3$
   $2x + y = 5$

7. $x + y = 6$
   $x - y = 8$

8. $x + y = 7$
   $x - y = 7$

9. $2x + 3y = 6$
   $x - y = -7$

10. $x + 8y = 12$
    $x - 2y = 2$

Graphing systems of inequalities

When you solve a system of equations by graphing, each equation has a graph that is a line, and the solution of the system is the point where the lines intersect. The solution of a system of inequalities is also the intersection of the solution sets of the individual inequalities, but the solution set of an inequality is not just a line, but a region, sometimes called a half-plane. The solution of the system of inequalities is the area where the two half-planes overlap.

To solve the system of inequalities $\begin{cases} x + 3y < 11 \\ y < 2x + 6 \end{cases}$, graph each of the inequalities, shading lightly, then locate the overlap and shade these shared values more prominently (see Figure 5.2).

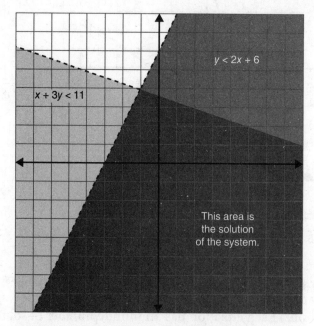

Figure 5.2 Solving a system of inequalities by graphing.

Graph each system of inequalities and indicate the solution set by shading.

1. $y \geq 2x - 3$
   $y > 3 - x$

2. $y < x + 5$
   $y \geq x + 3$

3. $y \geq 2x - 3$
   $y \geq 3 - 2x$

4. $y > x$
   $y > -x$
   $y < 4$

5. $y \leq x + 3$
   $y \leq 3 - x$
   $y \geq 0$

6. $3x - y > 0$
   $3x - 2y > 4$

7. $x - y > -5$
   $2x - y < -2$

8. $y \geq 2x + 2$
   $y < -x + 4$
   $x > -2$

9. $y > 2x + 1$
   $y \leq 2x - 3$

10. $y \geq -2x - 3$
    $y \leq x + 2$
    $y \geq 7x - 3$

Solving systems of equations by substitution

When one equation in a system expresses one variable as a function of the other, like $y = 3x - 1$ or $x = 4 + y$, it can be substituted into the other equation. That substitution produces an equation with only one variable, an equation that can be easily solved. To solve the system $\begin{cases} y = 3x - 1 \\ 2x + 5y = 12 \end{cases}$, replace the y in the bottom equation with $3x - 1$.

$$2x + 5y = 12$$
$$2x + 5(3x - 1) = 12$$
$$2x + 15x - 5 = 12$$
$$17x - 5 = 12$$
$$17x = 17$$
$$x = 1$$

Once you know the value of one variable, substitute that into one of the original equations to find the value of the other variable.

$$y = 3x - 1$$
$$y = 3(1) - 1$$
$$y = 2$$

Substitution is most convenient when one of the equations is given to you in $x =$ or $y =$ form, but you can use it anytime you can conveniently put one of the equations into that form. If trying to solve for x or for y in terms of the other variable is too difficult or gives you a messy substitution, combinations may be a better method.

Solve each system of equations by substitution.

1. $y = x$
    $x + y = 10$

2. $x + y = 12$
    $y = 2x$

3. $x - y = -6$
    $y = 3x$

4. $x + y = 57$
    $x = y + 3$

5. $x + 2y = 65$
    $y = x + 4$

6. $x - y = 46$
    $x = 7y - 2$

7. $3x - 4y = 5$
    $y = 2x - 5$

8. $5x + 7y = 73$
    $x - 2y = 1$

9. $5x - 3y = 49$
    $x - 4y = 3$

10. $y = 2x + 7$
    $y = 3x + 8$

Solving systems of equations by combination

If both equations are in standard form (or can easily be transformed to standard form), then the method of linear combinations, or elimination, allows you to add or subtract the equations in a way that will make one variable drop out. Make sure that both equations are in standard form and that like terms are aligned under one another before you begin.

Addition

If the coefficients of one variable are opposites, add the equations to eliminate that variable.

$$2x - 3y = -10$$
$$5x + 3y = -4$$
$$\overline{7x + 0y = -14}$$
$$x = -2$$

Solve for the remaining variable and substitute back into either equation to find the value of the second variable.

$$2(-2) - 3y = -10$$
$$-4 - 3y = -10$$
$$-3y = -6$$
$$y = 2$$

Subtraction

If the coefficients of one variable are identical, you can subtract the equations to eliminate that variable, or you can multiply one of the equations by -1 and then add.

$$4x + 2y = 12$$
$$\underline{3x + 2y = 7}$$
$$x + 0y = 5$$
$$x = 5$$

Solve for the remaining variable and substitute back into either equation to find the value of the second variable.

$$4(5) + 2y = 12$$
$$20 + 2y = 12$$
$$2y = -8$$
$$y = -4$$

EXERCISE

5·4

Solve each system of equations.

1. $x + y = 8$
   $x - y = 4$

2. $x + y = 17$
   $x - y = 3$

3. $2x + y = 9$
   $x - y = 3$

4. $3x + 2y = 23$
   $x - 2y = 5$

5. $5x + 12y = 16$
   $2x - 12y = -2$

6. $8x + y = 19$
   $2x + y = 7$

7. $7x + 5y = 21$
   $7x + 9y = 21$

8. $x + 9y = 93$
   $x + 4y = 43$

9. $x + y = 1$
   $2x + y = 9$

10. $x + 2y = 6$
    $3x - 2y = 6$

Combinations with multiplication

If the coefficients of the variable are neither identical nor opposite, you can still use combinations, but you need to do a little work first. Multiply one or both equations by constants to create equivalent equations with matching or opposite coefficients.

If you need to solve the system $\begin{cases} 2x+3y=17 \\ 5x-2y=14 \end{cases}$, you can multiply the top equation by 2 and the bottom equation by 3. That will turn the system into $\begin{cases} 4x+6y=34 \\ 15x-6y=42 \end{cases}$. This system has the same solution as the original, but in this version, you can eliminate y by adding the equations. Once the coefficients are identical or opposite, eliminate a variable by adding or subtracting, and then substitute back into one of the original equations.

Be careful to multiply through the entire equation by the constant. Multiplying the variable terms but not the constant is a common error. Checking your solution in both of the original equations will help you catch your errors.

EXERCISE

5·5

Solve each system of equations.

1. $3a+2b=15$
   $2a+b=8$

2. $3x+2y=23$
   $x-y=-9$

3. $8x+2y=16$
   $2x-y=7$

4. $5x-2y=29$
   $3x+4y=33$

5. $4x-3y=14$
   $3x+2y=19$

6. $2x+3y=4$
   $3x-8y=-9$

7. $3x-7y=30$
   $5x+y=12$

8. $7x-3y=1$
   $2x-y=1$

9. $2x+4y=5$
   $4x+5y=6$

10. $2x+3y=21$
    $4x+y=9$

Dependent and inconsistent systems

Not every system of equations has a unique solution. Since parallel lines never intersect, if the two equations produce parallel lines when graphed, the system has no solution. This occurs when the equations have the same slope but different y-intercepts. A system that has no solution is inconsistent. A system that has a unique solution is consistent.

If a system is made up of two equations that produce the same graph, every point on the line is a point of intersection, and so the system has infinitely many solutions. Such a system is dependent. You can recognize a dependent system because one equation will be a constant multiple of the other.

Label each system as consistent, inconsistent, or dependent.

1. $x + y = 3$
   $7x + 7y = 21$

2. $x + 2y = 7$
   $x + 2y = 9$

3. $x + y = 11$
   $x - y = -1$

4. $2x + y = 13$
   $8x + 4y = 51$

5. $2x - 5y = 3$
   $10x = 15 + 25y$

6. $7x + 3y = 24$
   $9x - 3y = 24$

7. $11x - 7y = 13$
   $14y = 46 - 22x$

8. $y = 2x - 3$
   $y = 2x + 4$

9. $y = 8 - 3x$
   $y = 8 - 5x$

10. $5x - 3y = 20$
    $x - 4 = \dfrac{3}{5}y$

Calculator notes #4

You can solve a system of linear equations by substitution or by elimination, and usually do so without too much trouble. Graphing is also a method for solving systems, but often not the most practical one, because drawing the graphs by hand can be time-consuming, and if the solutions are not integers, estimating the fractional part can be difficult. What looks like $3\frac{1}{2}$ on a hand-drawn graph could really be $3\frac{5}{8}$ or $3\frac{2}{5}$ and it would be hard to tell.

The graphing calculator gives you a quick way to graph a system, and will give you the solution, even if the values are not integers, accurate to several decimal places. This shouldn't replace your other methods of solving systems but it does give you a good way to verify the solution you found.

$$3x - 5y = 21$$
$$2x + 3y = -5$$

To solve the system above on your graphing calculator, you'll need to first isolate y in each equation. Remember that the calculator won't care if the expression is simplified, so don't fuss if it's not convenient to clean things up. Just remember to enclose the numerator in parentheses.

$$3x - 5y = 21 \qquad\qquad 2x + 3y = -5$$
$$-5y = 21 - 3x \qquad\qquad 3y = -5 - 2x$$
$$y = \frac{(21 - 3x)}{-5} \qquad\qquad y = \frac{(-5 - 2x)}{3}$$

- Enter the equations into the $\boxed{Y=}$ menu in Y1 and Y2.

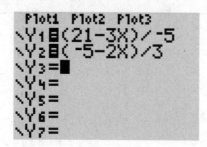

- Press $\boxed{ZOOM}$ and choose 6: ZStandard to set a standard viewing window and display the graphs. If you can't see the point of intersection of the lines, adjust your window.

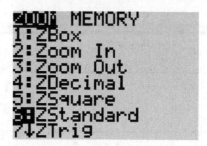

- With the point of intersection visible on screen, press $\boxed{2nd}$ $\boxed{TRACE}$, and choose 5:intersect.

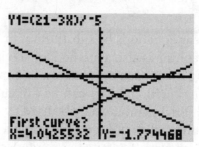

- The cursor appears on one line, and the question "First curve?" appears at the bottom of the screen. If you have only two lines, you can just press $\boxed{ENTER}$. (If you have more than two lines, use the up and down cursor keys to move to one of the lines of the system, then $\boxed{ENTER}$.)
- The cursor jumps to a different line, and question "Second curve?" appears. As long as it's the other line of the system, just press $\boxed{ENTER}$.
- The question changes to "Guess?" For linear systems, you can just press $\boxed{ENTER}$. If you wish, you can move your cursor close to the point of intersection before you press $\boxed{ENTER}$. (Later, when you solve systems in which graphs intersect more than once, the guess option lets you move the cursor to the intersection you want.)
- The solution of the system displays at the bottom of the screen.

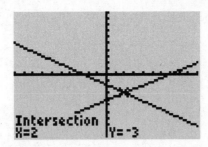

In addition to allowing you to find the solution of a system, the graph also lets you see when a system is inconsistent or dependent. If the lines are parallel, there is no point of intersection to find, so the system has no solution and is inconsistent. If the graph appears to be only one line, go to [Y=], cursor down to Y2, cursor to the left and press [ENTER] to change the symbol. Press [GRAPH] and you show be able to see if the second line traces right on top of the first. If it does, the system is dependent.

Powers and polynomials ·6·

An exponent is a symbol used to show repeated multiplication. The product $5 \times 5 \times 5$ is written 5^3 to show that 5 is used as a factor 3 times. The expression x^4 means $x \cdot x \cdot x \cdot x$, or the product obtained by using x as a factor 4 times. The number (or variable) that is multiplied is called the base, the little number that tells how many times to use it is the exponent, and together, as in 5^3 or x^4, they form a power.

Rules for exponents

To multiply powers of the same base, keep the base and add the exponents. If you write out in long form what the powers mean, you can see that the result of multiplying powers of the same base is another power of that base, and the new exponent can be found by adding the exponents in the problem.

$$a^2 \cdot a^3 = (a \cdot a)(a \cdot a \cdot a) = a \cdot a \cdot a \cdot a \cdot a = a^5 = a^{2+3}$$

If the powers have different bases, there's really not much you can do. $4^2 \cdot 3^3$ is the product of two 4s and three 3s. It's not five of anything. You could evaluate 4^2 and evaluate 3^3 and multiply the results, but that won't work when the bases are variables.

To divide powers of the same base, keep the base and subtract the exponents. Again, if you write out the long form of the problem and cancel, you'll see that the result is a power of the same base, with an exponent that's the difference between the two exponents.

$$\frac{t^7}{t^3} = \frac{t \cdot t \cdot t \cdot t \cdot \cancel{t} \cdot \cancel{t} \cdot \cancel{t}}{\cancel{t} \cdot \cancel{t} \cdot \cancel{t}} = t^4 = t^{7-3}$$

To raise a power to a power, keep the base and multiply the exponents. If you say you want to square a power, for example, you're saying you want to multiply it by itself, to use it as a factor twice. That turns it into a multiplication problem, and you could follow the rule for multiplication. This rule is just a shortcut.

$$(b^3)^2 = (b^3)(b^3) = b^{3+3} = b^6 = b^{2 \times 3}$$

Special exponents

If you follow the rule for dividing powers to evaluate $\frac{x^5}{x^5}$, you'll conclude that $\frac{x^5}{x^5} = x^0$. You know

from arithmetic, however, that any number divided by itself equals 1, so $\frac{x^5}{x^5} = 1$. Put those two

ideas together and you get a simple rule: any non-0 number to the 0 power is 1. If $a \neq 0$, $a^0 = 1$.
Notice that the rule applies only to non-0 numbers. If you tried to do 0^0, you'd be torn between
"any number to the 0 power is 1" and "0 to any power is 0." The definition of the 0 power comes
from dividing, and because division by 0 is undefined, 0^0 is indeterminate.

Applying the division rule to a problem that has a larger power in the denominator than in the

numerator leads to another definition. According to the rules for exponents, $\frac{x^5}{x^6} = x^{-1}$ and $\frac{y^3}{y^7} = y^{-4}$.

To understand what these negative exponents mean, write out the problems in their long form.

$$\frac{x^5}{x^6} = \frac{\cancel{x} \cdot \cancel{x} \cdot \cancel{x} \cdot \cancel{x} \cdot \cancel{x}}{\cancel{x} \cdot \cancel{x} \cdot \cancel{x} \cdot \cancel{x} \cdot \cancel{x} \cdot x} = \frac{1}{x} = x^{-1}$$

$$\frac{y^3}{y^7} = \frac{\cancel{y} \cdot \cancel{y} \cdot \cancel{y}}{\cancel{y} \cdot \cancel{y} \cdot \cancel{y} \cdot y \cdot y \cdot y \cdot y} = \frac{1}{y^4} = y^{-4}$$

Simplify each expression.

1. $x^4 \cdot x^7$

2. $y^5 \cdot y$

3. $6x^3 \cdot x^3$

4. $3x^5 \cdot 7x^5$

5. $\dfrac{x^8}{x^2}$

6. $\dfrac{t^3}{t^4}$

7. $\dfrac{y^8}{y}$

8. $\dfrac{x^4}{x^9}$

9. $\dfrac{x^4 \cdot x^7}{x^9}$

10. $(y^7)^3$

11. $\dfrac{x^4 \cdot x^{-2}}{x^2}$

12. $(x^2)^5$

13. $\dfrac{x^3 \cdot x^8}{(x^3)^3}$

14. $\dfrac{t^2(t^5)^3}{(t^9)^2}$

15. $\dfrac{x^6 \cdot x^9}{x^5} \cdot \left(\dfrac{x^3}{x^2}\right)^{-4}$

More rules

Combining the basic rules for exponents with the associative and distributive properties produces advanced rules for the power of a product and the power of a quotient.

Power of a product

When a product of two or more factors is raised to a power, the associative and commutative properties allow us to find a shortcut. Write out the power, rearrange, and regroup, and you'll see that each factor in the product is raised to the power.

$$(4x^2y^5)^3 = (4x^2y^5)(4x^2y^5)(4x^2y^5)$$
$$= 4 \cdot 4 \cdot 4 \cdot x^2 \cdot x^2 \cdot x^2 \cdot y^5 \cdot y^5 \cdot y^5$$
$$= 4^3(x^2)^3(y^5)^3$$

When a product is raised to a power, each factor is raised to that power.

Power of a quotient

When a quotient is raised to a power, both the numerator and the denominator are raised to that power.

$$\left(\frac{y}{x^5}\right)^2 = \frac{y}{x^5} \cdot \frac{y}{x^5} = \frac{y^2}{(x^5)^2}$$

You can combine the power of a product rule and the power of a quotient rule to handle more complicated expressions, but don't misapply them. These rules don't apply to sums or differences.

EXERCISE
6·2

Simplify each expression.

1. $(2x^5)^2$

2. $(-2x^3)^3$

3. $(5a^2)(2a^3)^2$

4. $(-x^2)(3xy^5)^3$

5. $(3b)^2(2b^3)^3$

6. $\left(\dfrac{x^5}{3}\right)^2$

7. $\left(\dfrac{2x^3}{5x}\right)^2$

8. $\left(\dfrac{4t^3}{3t^5}\right)^4$

9. $\left(\dfrac{-4x^3y^5}{3x^2y}\right)^3$

10. $\left(\dfrac{3x^2}{2y}\right)^4$

Monomials and polynomials

A monomial is a single term. Monomials involve only multiplication of real numbers, variables, and positive integer powers of variables. The expressions -3, x^2, and $\frac{2}{5}t^7$ are all monomials, but $\frac{4}{y^3}$

is not because it involves division by a variable, and $\sqrt{x}$ is not because it can't be written as an integer power of x.

When monomials are combined by addition or subtraction, they form polynomials. A polynomial with two terms is a binomial, and one with three terms is a trinomial. For four or more terms, we use the general term polynomial.

Degree of a polynomial

The degree of a monomial containing one variable is the power to which the variable is raised. The degree of $3x^5$ is 5. The degree of x is 1, and the degree of any constant is 0. If a monomial contains more than one variable, for example, $-6x^2 y^3$, its degree is the sum of the powers. The expression $-6x^2 y^3$ is a fifth-degree monomial. The degree of a polynomial is the degree of its highest-power monomial. The degree of $6x^4 - 3x^2 + 12$ is 4.

Standard form

A polynomial is in standard form when its terms are arranged in order from highest degree to lowest degree or lowest to highest degree. The polynomial $6x^4 - 3x^2 + 12$ is in standard form, but $-7t^5 + 8t^2 - 3t^7 + 2t - 1$ is not.

EXERCISE 6·3

Put the polynomials in standard form and give the degree. If the expression is not a polynomial, explain why.

1. $5x + 3x^2 - 7 + 2x^3$

2. $t^7 - 1 + 5t^{12} + 8t^2 - 9t$

3. $5y^6 - 2y^3 + 8 - 12y^{11}$

4. $\sqrt{t^2 - 4t + 5}$

5. $2x^5 - 4x^3 + 3x$

6. $4 - 3z^7 + 8z - 4z^2$

7. $7 - 3w + w^5 - 9w^3$

8. $b^2 - 3b - 4 - b^4$

9. $\dfrac{1}{x - 4}$

10. $6 - 7y^3 + 8y^2 - 4y$

Adding and subtracting polynomials

To add polynomials, follow the rules for combining like terms. If the terms match in both variable and power, add the coefficients. Keep the variable portion unchanged. Unlike terms cannot be combined. You will sometimes see polynomials enclosed by parentheses, as in $(5x^3 - 9x^2 + 7x - 4) + (2x^4 - 8x^3 - 5x^2 + 3)$. These parentheses are just to define the polynomials and can be dropped when adding. Rearrange the terms to bring like terms together and simplify.

$$(5x^3 - 9x^2 + 7x - 4) + (2x^4 - 8x^3 - 5x^2 + 3)$$
$$= 5x^3 - 9x^2 + 7x - 4 + 2x^4 - 8x^3 - 5x^2 + 3$$
$$= 2x^4 + 5x^3 - 8x^3 - 9x^2 - 5x^2 + 7x - 4 + 3$$
$$= 2x^4 - 3x^3 - 14x^2 + 7x - 1$$

When you are subtracting polynomials, the parentheses have significance. To subtract polynomials, it is possible to subtract term by term. For example, the subtraction $(4y^3 + 3y^2 + 7) - (-2y^3 - 8y^2 + y - 2)$ can be thought of as

$$[4y^3 - (-2y^3)] + [3y^2 - (-8y^2)] + (0y - y) + [7 - (-2)]$$

It is usually simpler, however, to treat subtraction as adding the opposite. To subtract $(4y^3 + 3y^2 + 7) - (-2y^3 - 8y^2 + y - 2)$, think of it as $(4y^3 + 3y^2 + 7)$ plus the opposite of $(-2y^3 - 8y^2 + y - 2)$. The opposite of $(-2y^3 - 8y^2 + y - 2)$ is $(2y^3 + 8y^2 - y + 2)$. In the problem $(4y^3 + 3y^2 + 7) - (-2y^3 - 8y^2 + y - 2)$, imagine that the subtraction sign is distributed to all the terms in the second set of parentheses; then drop the parentheses and add.

$$(4y^3 + 3y^2 + 7) - (-2y^3 - 8y^2 + y - 2)$$
$$= (4y^3 + 3y^2 + 7) + (2y^3 + 8y^2 - y + 2)$$
$$= 4y^3 + 3y^2 + 7 + 2y^3 + 8y^2 - y + 2$$
$$= 4y^3 + 2y^3 + 3y^2 + 8y^2 - y + 7 + 2$$
$$= 6y^3 + 11y^2 - y + 9$$

EXERCISE

6·4

Add or subtract the polynomials as indicated and give your answers in simplest form.

1. $(13w^2 - 9w + 8) + (w^2 - 9)$

2. $(a^2 - 7a + 5) + (a^2 + 2a - 9)$

3. $(-13x^2 + 43x - 27) + (4x^2 - 2x + 3)$

4. $(3y^2 - 17y + 34) + (-7y^2 + 14y - 2)$

5. $(1 - 2b + b^2) - (-3 - b - 3b^2)$

6. $(2b^2 + 6b - 5) + (8 - 9b + 2b^2)$

7. $(9x^2 - 7x + 5) - (-2x^2 + 6x + 3)$

8. $(2x^2 - 3x) - (4x^2 + 4x - 2)$

9. $(2x^2 - x + 5) - (5x^2 - 2x + 3)$

10. $(5x^2 - 7x + 2) - (3x^2 + 9x - 1)$

Multiplying polynomials

All polynomial multiplication is built on multiplying monomials, but we have different rules for polynomials of different sizes to make the work more efficient.

Multiplying monomials

To multiply two monomials, first multiply the coefficients. If the variables are the same, use rules for exponents to simplify. If the variables are different, just write them side by side. The product $(-3x^5y^2)(2x^2y^3) = (-3 \cdot 2)(x^5 \cdot x^2)(y^2 \cdot y^3) = -6x^7y^5$, but the product $(5a^2)(2b^3) = (5 \cdot 2)(a^2)(b^3) = 10a^2b^3$.

EXERCISE
6·5

Multiply the monomials and give your answers in simplest form.

1. $(-3b^2)(2b^5)$

2. $(-6xy^2)(-5x^3y^2)$

3. $(9x^2yz^5)(-4x^3yz^5)$

4. $(-ab^2)(3ac^3)$

5. $(5ab)(8a^2)$

6. $-x^2(3xy)(-6x^3y^2)$

7. $-6w^3(2wx^2)(3wx^4)$

8. $(2x^3)^2$

9. $(5b^2)(2b^3)^2$

10. $(-t^3)(3rt^2)^3$

Fill in the blanks with the missing monomial factor.

11. $(2x^3)(\underline{\quad}) = -6x^5$

12. $(-3b^2)(\underline{\quad}) = 12b^7$

13. $(\underline{\quad})(3x^2y) = -15x^6y^3$

14. $(\underline{\quad})(-2z^4) = 6x^2z^5$

15. $(6xy^2)(\underline{\quad}) = -3x^3y^3$

Multiplying with the distributive property

To multiply a monomial times a larger polynomial, use the distributive property.

$$-2x^3(4x^5 - 6x^4 + 5x^3 - 7x^2 + 8x - 2)$$
$$= (-2x^3 \cdot 4x^5) + (-2x^3 \cdot -6x^4) + (-2x^3 \cdot 5x^3) + (-2x^3 \cdot -7x^2) + (-2x^3 \cdot 8x) + (-2x^3 \cdot -2)$$

Then follow the rules for multiplying monomials.

$$(-2x^3 \cdot 4x^5) + (-2x^3 \cdot -6x^4) + (-2x^3 \cdot 5x^3) + (-2x^3 \cdot -7x^2) + (-2x^3 \cdot 8x) + (-2x^3 \cdot -2)$$
$$= -8x^8 + 12x^7 - 10x^6 + 14x^5 - 16x^4 + 4x^3$$

EXERCISE
6·6

Multiply the polynomials and give your answers in simplest form.

1. $5a(2a^2 + 3a)$

2. $-2x^2(x^2 - 3x - 2)$

3. $2y^2(11y^2 - 3y + 5)$

4. $-3b^3(2b^2 - 3b + 4)$

5. $xy(3x^2 + 5xy - 2y^2)$

6. $5x^2y(5x^2 - 7xy + y^2)$

7. $8x(x + 2y - 3z)$

8. $-5ab(a^2 - b^3)$

9. $\frac{1}{2}x^3(8x^7 - 6x^5 + 10x^4 - 2x^2 + 14x - 20)$

10. $-3a^4b^3c^2(-3a^2b + 2bc - 7a^5c^4)$

Fill in the blanks with the missing factor.

11. $(\underline{})(x+1)=3x+3$

12. $a(\underline{})=ab-5a$

13. $(\underline{})(2x-y)=8x-4y$

14. $7x(\underline{})=7x+49x^2$

15. $(\underline{})(2a+b)=4a^2b+2ab^2$

Multiplying binomials

Multiplication of two binomials is accomplished by repeated application of the distributive property, but there is a convenient shortcut, known by the acronym FOIL.

The distributive rule

To multiply two binomials by distributing, treat the first binomial as the multiplier and distribute it to both terms of the second binomial.

$$(x+5)(x-3)=x(x+5)-3(x+5)$$

That gives you two smaller distributive multiplications. Distribute x to both terms of $x+5$, distribute -3 to both terms of $x+5$, and combine like terms.

$$x(x+5)-3(x+5)=x^2+5x-3x-15=x^2+2x-15$$

The FOIL rule

FOIL stands for First, Outer, Inner, Last and is a reminder of the four multiplications that must be performed to successfully multiply two binomials.

First: $(\underline{2x}-3)(\underline{5x}+7)=(\underline{2x})(\underline{5x})$

Outer: $(\underline{2x}-3)(5x\underline{+7})=(2x)(5x)+(\underline{2x})(\underline{7})$

Inner: $(2x\underline{-3})(\underline{5x}+7)=(2x)(5x)+(2x)(7)+(\underline{-3})(\underline{5x})$

Last: $(2x\underline{-3})(5x\underline{+7})=(2x)(5x)+(2x)(7)+(-3)(5x)+(\underline{-3})(\underline{7})$

You will often find that there are like terms that can be combined after the four multiplications are performed.

$$(2x-3)(5x+7)=(2x)(5x)+(2x)(7)+(-3)(5x)+(-3)(7)$$
$$=10x^2+14x-15x-21$$
$$=10x^2-x-21$$

When the binomials you're multiplying are the sum and difference of the same two terms, like $(x+5)$ and $(x-5)$, the inner and outer terms will add to 0, leaving you with a difference of squares.

$$(x+5)(x-5)=x^2-5x+5x-25$$
$$=x^2-25$$

EXERCISE 6·7

Use the FOIL rule to multiply the binomials. Give your answers in simplest form.

1. $(x+8)(x+2)$

2. $(y-4)(y-9)$

3. $(t-2)(t+6)$

4. $(2x+8)(x-3)$

5. $(y-9)(3y+1)$

6. $(5x-6)(3x+4)$

7. $(6x-1)(x+5)$

8. $(1-3b)(5+2b)$

9. $(3x-7)(2x+5)$

10. $(5-2x)(5x-2)$

Try to predict the product of each pair of binomials without actually multiplying; then check with the FOIL rule.

11. $(x-4)(x+4)$

12. $(x+3)(x-3)$

13. $(2x-1)(2x+1)$

14. $(3x+5)(3x-5)$

15. $(7-3x)(7+3x)$

Fill in the blank with the missing term. Check your answer by FOILing.

16. $(x+3)(x+\underline{})=x^2+5x+6$

17. $(x-7)(x-\underline{})=x^2-9x+14$

18. $(2a+1)(a+\underline{})=2a^2+9a+4$

19. $(3x-2)(x-\underline{})=3x^2-17x+10$

20. $(2t+3)(3t-\underline{})=6t^2-t-15$

Multiplying larger polynomials

When one of the polynomials to be multiplied has more than two terms, it may be convenient to place them one under another, usually with the longer one on top and shorter one on the bottom, and multiply each term in the bottom polynomial by each term in the top polynomial, arranging like terms under one another in the result for easy combination. This is the same algorithm you learned for multiplying numbers with more than two digits.

$$
\begin{array}{r}
2x^2 + 3x - 4 \\
3x - 1 \\
\hline
-2x^2 - 3x + 4 \qquad \text{(−1 times each term)} \\
6x^3 + 9x^2 - 12x \qquad \text{(3x times each term)} \\
\hline
6x^3 + 7x^2 - 15x + 4 \qquad \text{(Add like terms)}
\end{array}
$$

Multiply the polynomials in a vertical format. Keep like terms aligned and give your answer in simplest form.

1. $(a+7)(2a^2+5a+3)$

2. $(2b+3)(3b^2+2b+5)$

3. $(c-8)(4c^2-7c-2)$

4. $(2x-1)(4x^3-7x^2+5x-7)$

5. $(y+4)(y^2-5y+1)$

6. $(x-2)(x^2-4x+4)$

7. $(t-2)(t^2+2t+4)$

8. $(x+1)(x^2-x+1)$

9. $(x^2-4)(x^2+4x+4)$

10. $(y^2-3y+5)(2y^2+5y-3)$

Dividing polynomials

Polynomial division is built on dividing monomials, but there are systems to help organize larger problems.

Dividing by a monomial

To divide a monomial by a monomial, divide the coefficients and use the rules for exponents to simplify like variables.

$$\frac{-4x^3y^2}{2x^2y} = \frac{-4}{2}\left(\frac{x^3}{x^2}\right)\left(\frac{y^2}{y}\right) = -2xy$$

To divide a larger polynomial by a monomial, divide each term of the larger polynomial by the monomial divisor.

$$\frac{12t^5-8t^3+5t^2+2t}{2t} = \frac{12t^5}{2t} - \frac{8t^3}{2t} + \frac{5t^2}{2t} + \frac{2t}{2t} = 6t^4 - 4t^2 + \frac{5}{2}t + 1$$

Divide the polynomials and give your answer in simplest form.

1. $(24c^7d^2) \div (3c^4)$

2. $(-35d^6) \div (-7d^2)$

3. $(-52x^9) \div (-13x^2)$

4. $(20y^{12}) \div (4y^8)$

5. $(24x^5y) \div (-6x^3)$

6. $(15x^2) \div (15x^2)$

7. $(-3.9t^6) \div (1.3t^4)$

8. $(-18q^7r^3) \div (q^7r)$

9. $(8x^4-9x^7) \div (x^3)$

10. $(15y^5-21y^7) \div (3y^2)$

11. $(5x^2+15x) \div (5x)$

12. $(-9y^3+13y^4) \div (-y^3)$

13. $(56z^{10}-49z^9+42z^8-35z^7) \div (7z^2)$

14. $(-9x^4y^3+27x^3y^4-81x^2y^5) \div (-3xy)$

15. $(15x^3-5x^2+20x) \div (5x)$

Long division

Long division of polynomials is modeled on the algorithm for long division that you learned in arithmetic. It can be used to divide by a monomial, but it is more commonly used when the divisor is a binomial or a larger polynomial.

Arrange the dividend and the divisor in standard form, highest power to lowest, and insert 0s for any missing powers to make it easier to line up like terms. Divide the first term of the dividend by the first term of the divisor and place the result as the first term of the quotient. Multiply the entire divisor by the term you just placed in the quotient, aligning like terms under the dividend. Subtract, and bring down any remaining terms in the dividend.

$$
\begin{array}{r}
3x^2 \\
2x+4\overline{)\,6x^3+\ 0x^2+8x+10} \\
\underline{6x^3+12x^2\quad\downarrow\quad\downarrow} \\
-12x^2+8x+10
\end{array}
$$

($6x^3 \div 2x = 3x^2$ goes to the quotient.)

(Multiply $3x^2$ times $2x+4$, subtract, and bring down.)

($-12x^2+8x+10$ is the new dividend.)

Repeat those steps, but use this new expression formed by subtracting and bringing down as your dividend.

$$
\begin{array}{r}
3x^2-6x+16 \\
2x+4\overline{)\,6x^3+0x^2+8x+10} \\
\underline{6x^3+12x^2\quad\downarrow\quad\downarrow} \\
-12x^2+8x+10 \\
\underline{-12x^2-24x\quad\downarrow} \\
32x+10 \\
\underline{32x+64} \\
-54
\end{array}
$$

($-12x^2 \div 2x = -6x$)

($-6x$ times $2x+4$)

($32x \div 2x = 16$)

(16 times $2x+4$)

You can express the remainder as a fraction by putting the remainder as the numerator of the fraction and the divisor as the denominator. The division problem $(6x^3+8x+10)\div(2x+4)$ is equal to $3x^2-6x+16+\dfrac{-54}{2x+4}$.

EXERCISE 6·10

Divide using long division.

1. $(x^2-15x+56)\div(x-8)$

2. $(y^2-y-20)\div(y-5)$

3. $(6x^2+5x-6)\div(3x-2)$

4. $(84x^4-3x^2-45)\div(12x^2-9)$

5. $(9x^2-42x+45)\div(3x-8)$

6. $(2a^2+7a+5)\div(a+3)$

7. $(2b^2-7b+3)\div(b-3)$

8. $(12x^3+17x^2-20x-20)\div(3x+5)$

9. $(x^4+8x^2+12)\div(x^2+2)$

10. $(8y^3+1)\div(2y+1)$

Calculator notes #5

When you're working with polynomials, you have variables and powers, sometimes more than one variable, no specific value to plug in for the variable, and you need to add, subtract, multiply, and divide. How is the calculator going to help?

Operations with polynomials (and operations with rational expressions) are not going to be taken over by your calculator, unless your calculator is equipped with CAS, but there is a simple way to use the graphing calculator to check your work when you're operating on polynomials in one variable. Let's say you need to simplify the expression below.

$$3x^2\left(2x^2 - 5x + 3\right) - \left(2x - 7\right)\left(x^2 + 4\right)$$

You carefully apply the distributive property and the FOIL rule. You're alert to the fact that subtracting can lead to sign errors. You combine like terms and only like terms, and when you're done, you have $6x^4 - 17x^3 + 16x^2 - 8x + 28$. But your friend has $4x^3 - 8x^2 + x + 28$. You don't want to start over. What to do?

- Go to the $\boxed{Y=}$ menu, and clear any old equations.
- Enter the original problem in the Y1 = slot. Graph. Adjust the window, if necessary, so that you can see a good bit of the graph.

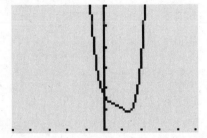

- Enter your result in the Y2 = slot. Move your cursor all the way to the left and $\boxed{ENTER}$. The slanted line will change to a heavier slanted line. If you press again, it will change again and again until you come round to where you started. Choose either the heavy line or one of the little "bubbles" and then graph.

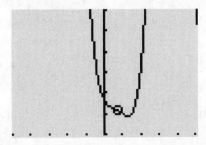

- If the graph of your answer follows right along the graph of the question, you have an equivalent expression. You don't have any mistakes, but always check to make sure you simplified completely. The calculator won't tell you that.

Factoring

7

When you learn to multiply, you learn that each of the numbers you multiply are called factors, and the result of that multiplication is a product. There are many different sets of factors that produce the same product. The factors 2 and 24 yield the product 48 but so do the factors 3 and 16, and the factors 2, 3, and 8, and many others.

You can talk about factor pairs, two numbers that multiply to a particular product, and discover that a product may have multiple factor pairs. The factor pairs for 12 are 1 and 12, 2 and 6, and 3 and 4. You can also talk about the prime factorization of a number, which reduces it to a product of the smallest possible integer factors, all prime numbers. Prime factorization of large numbers, like 108, is easier to write with exponents: $108 = 2 \times 2 \times 3 \times 3 \times 3 = 2^2 \times 3^3$.

Factoring is the process of reexpressing a quantity as the product of two or more simpler quantities, called factors. You can factor 35 by writing it as 5×7, and you can factor a monomial like $-3x^2 y^3$ by writing out $-3 \cdot x \cdot x \cdot y \cdot y \cdot y$. Factoring polynomials is a little more complicated, but a few rules will cover most situations.

Greatest common monomial factor

You will often be able to find several different factor pairs for a monomial. You know this is true when you factor a constant, but it's also true for a single term involving a variable. The constant 24 could be factored as 1×24, 2×12, 3×8, or 4×6. The monomial $48x^5$ could be written as $48 \cdot x^5$, $12x^2 \cdot 4x^3$, $3x \cdot 16x^4$, and more.

Identifying the GCF

The greatest common monomial factor of a polynomial is the largest monomial that is a factor of every term. In this context, *largest* means the "largest coefficient and the highest power of the variable." The polynomial $12x^5 y + 8x^4 y^2 - 10x^3 y^3$ has a greatest common factor of $2x^3 y$ because 2 is the largest integer that divides all three coefficients, x^3 is the largest power of x present in all terms, and y is the largest power of y present in all terms. Notice that the largest power of a variable contained in all the terms is the smallest one you see. The largest power of x in $12x^5 y + 8x^4 y^2 - 10x^3 y^3$ is x^3, which is the smallest power of x in any of the terms. You can leave the extra variables behind, but you can't factor out more than the monomial has.

Identify the greatest monomial that is a factor of each term of the polynomial.

1. $4x^3 - 2x^2$

2. $25r^2 + 15r$

3. $22t^3 + 33$

4. $6x^3 - 8x^2 + 14x$

5. $2ax - 4a^2x^2 + 9ax^2$

6. $16t^5 - 24t^3 + 8t^4$

7. $az^2 - 2az + 10a^2z^3 - 4az^5$

8. $3x^2y^2 - 27y^2$

9. $2ax^2 - 10a^2x + 8a^3$

10. $6ab + 9bc - 12ac$

Writing the factored form

Factoring out the greatest common factor is applying the distributive property in reverse. The distributive property tells you that to multiply a monomial times a polynomial, you multiply the monomial by each term of the polynomial. To multiply $3x^3(x^2 - 7x + 4)$, multiply $3x^3$ times x^2, then $3x^3$ times $-7x$, and $3x^3$ times 4.

$$3x^3(x^2 - 7x + 4) = 3x^3 \cdot x^2 + 3x^3(-7x) + 3x^3 \cdot 4$$
$$= 3x^5 - 21x^4 + 12x^3$$

Each term of the new polynomial can be simplified, but because the trinomial in the original problem had no like terms, there should not be any like terms in the answer.

Factoring out a greatest common factor means starting with the product, working out what the monomial multiplier would be, and rewriting the polynomial as a product. To do this reversal on $3x^5 - 21x^4 + 12x^3$, first ask what is the largest integer that will divide all three coefficients. The largest divisor of 3, −21, and 12 is 3, so the greatest common factor has a coefficient of 3. Then look at the variables. What's the largest power of x that is a factor of x^5, x^4, and x^3? That will be the lowest power of the three, so x^3. The GCF is $3x^3$.

To find the other factor, the polynomial in parentheses, divide each term of the original polynomial by the GCF: $\dfrac{3x^5}{3x^3} - \dfrac{21x^4}{3x^3} + \dfrac{12x^3}{3x^3} = x^2 - 7x + 4$. To write the factored form, bring the common factor to the front, and in parentheses, show the other factors.

$$3x^5 - 21x^4 + 12x^3 = 3x^3(x^2 - 7x + 4).$$

Here's another example. The greatest common factor of the terms in $12x^5y + 8x^4y^2 - 10x^3y^3$ is $2x^3y$. Divide each term by the GCF.

$$\frac{12x^5y}{2x^3y} + \frac{8x^4y^2}{2x^3y} - \frac{10x^3y^3}{2x^3y} = 6x^2 + 4xy - 5y^2$$

Write as a product of the GCF and this simpler polynomial.

$$12x^5y + 8x^4y^2 - 10x^3y^3 = 2x^3y(6x^2 + 4xy - 5y^2)$$

If the greatest common factor is actually a term of the polynomial, remember to express that term as the common factor times 1 and have a 1 in the parentheses for that position.

$$6x^3 - 12x^2 + 3x = \underline{3x} \cdot 2x^2 - \underline{3x} \cdot 4x + \underline{3x} \cdot 1$$
$$= 3x(2x^2 - 4x + 1)$$

EXERCISE 7·2

Factor each polynomial by factoring out the greatest common monomial factor.

1. $y^2 - 15y$

2. $3b^2 - 6b$

3. $32a^2b + 40ab$

4. $5y^2 + 15y + 20$

5. $x^8y^4 - x^4y^7 + x^3y^5$

6. $-a^2 - a^3 + 2a^4$

7. $25x^4 - 50x^5 + 125x^7$

8. $8r^2 + 24rt + 16r$

9. $16x^2y^2 - 48x^3y$

10. $3x^2y + 6xy + 15x^2y^2$

Factoring $x^2 + bx + c$

A trinomial of the form $x^2 + bx + c$, if it is not prime, can be factored to the product of two binomials. Just as we say a number is a prime number if its only factors are itself and 1, a polynomial is prime if it's not factorable. If you factor $x^2 + bx + c$, it will factor to something of the form $(x + r)(x + t)$. When you factored out a greatest common monomial factor, you were reversing the type of multiplication you did with the distributive property. When you factor $x^2 + bx + c$, you reverse a multiplication done with the FOIL rule.

$$(x + r)(x + t) = x \cdot x + x \cdot t + r \cdot x + r \cdot t$$
$$= x^2 + (r + t)x + rt$$

The product of r and t will equal the constant term, c, of the trinomial and the sum of r and t will produce the middle coefficient b. To factor $x^2 + 5x + 6$, you need to find a pair of integers that add to 5 and multiply to 6. For small numbers, this isn't difficult: $x^2 + 5x + 6 = (x + 3)(x + 2)$. You can check your factors by multiplying, using the FOIL rule.

If the middle term of the trinomial is negative, as in $x^2 - 8x + 12$, the process of factoring is the same, except that the binomials have negative signs. The factors of $x^2 - 8x + 12$ are $(x - 6)(x - 2)$ because $-6 \cdot (-2) = 12$ and $-6 + (-2) = -8$.

When the constant term is negative, however, it's a signal that one factor is x plus a constant and one is x minus a constant, and that causes a slight change in your search for factors. If the constant term of the trinomial is negative, you still want two integers that multiply to that number, but you need to remember that adding a positive and a negative will look like subtracting.

To factor $x^2 - 9x - 90$, look for factors of -90, which are just factors of 90 but you'll make one negative. The factors of 90 are 1×90, 2×45, 3×30, 5×18, 6×15, and 9×10. None of those are going to add to 9 or -9, but that's OK. You're actually looking for a pair that subtracts to 9, and 6×15

fits the bill. Set up $(x \quad 6)(x \quad 15)$ and then think about the result you want. You want to get -9 as the coefficient of x, so you want a negative sign on the larger number, 15, and a positive sign on the smaller one, 6.

$$x^2 - 9x - 90 = (x + 6)(x - 15)$$

EXERCISE 7·3

Factor each polynomial. If the polynomial does not factor, write Prime.

1. $x^2 + 12x + 35$

2. $x^2 + 11x + 28$

3. $x^2 - 8x + 15$

4. $x^2 - 7x + 12$

5. $x^2 + x - 20$

6. $x^2 - 2x - 3$

7. $x^2 - 11x + 18$

8. $x^2 - 9x - 22$

9. $x^2 + 10x - 39$

10. $x^2 + 12x + 32$

Factoring $ax^2 + bx + c$

A trinomial of the form $ax^2 + bx + c$ is created by a FOIL multiplication, but the two binomials are not just x plus a constant. One or both of the x's have a coefficient other than one, and as you can see below, that introduces some complications.

$$(px + r)(qx + t) = px \cdot qx + px \cdot t + r \cdot qx + r \cdot t$$
$$= (pq)x^2 + (pt + qr)x + rt$$

When you're asked to factor a trinomial in which the coefficient of the x^2 term is a number other than 1, the factors of that x^2 coefficient a, as well as the factors of the constant c, affect the middle term. The good news is the constant term, c, is still the product of r and t, the constants in the binomial, but the factoring process becomes more a case of trial and error.

To get through that trial-and-error process as efficiently as possible, make a list of the factors of the lead coefficient and a list of the factors of the constant. In theory, you want to check all the possibilities, in all possible combinations, to see whether the "Inner" and the "Outer" from the FOIL rule will combine to form the middle term you need. You need to be very organized, going down your lists in order and crossing factors off when you're sure they don't work.

To make things a little easier, you can make a chart of the possible products and look for a pair that will add or subtract to the coefficient of the middle term. To factor $6x^2 + 5x - 21$, first notice that the factors of $6x^2$ are $1x \cdot 6x$ or $2x \cdot 3x$ and factors of 21 are $1 \cdot 21$ or $3 \cdot 7$. Put the factors of $6x^2$ down the side of the chart and the factors of 21 across the top. Keep pairs together. Then fill in the boxes with the products of the number at the beginning of the row and the top of the column.

	1	21		3	7
$1x$	$1x$	$21x$		$3x$	$7x$
$6x$	$6x$	$126x$		$18x$	$42x$
$2x$	$2x$	$42x$		$6x$	$14x$
$3x$	$3x$	$63x$		$9x$	$21x$

Look at the numbers on the diagonals of small squares. There's no way you can get a middle term of $5x$ from $1x$ and $126x$. The other diagonal in that square will give you $6x$ and $21x$, which could add to $27x$ or subtract to $15x$, but you're looking for $5x$, so keep looking. In the bottom-right square, 14 and 9 can subtract to 5, so try the numbers at the beginning of those rows and the top of those columns. You want to use $2x$ and $3x$ as the factors of $6x^2$, and 3 and 7 as the factors of 21, and place them so the inner and outer are $9x$ and $14x$.

$$6x^2 + 5x - 21 = (2x \quad 3)(3x \quad 7)$$

When you try to place the $+$ and the $-$, don't just look at 3 and 7. Instead, look at the inner and the outer. The inner is $9x$ and the outer is $14x$. You want the larger one, $14x$, to be positive, so put the $+$ on the 7 and the $-$ on the 3.

$$6x^2 + 5x - 21 = (2x - 3)(3x + 7)$$

Another strategy for locating the correct factors often sounds like it's going to make the problem worse, but if you use it wisely, it can help. It tells you to look at the trinomial and multiply the constant term by the coefficient of x^2. In the example above, that would mean 6×21, and you may feel like bigger numbers are not helping, but you do want to look for factors of 126. Instead of multiplying 6×21 to get 126, write the product 6×21 as a prime factorization: $6 \times 21 = 2 \times 3 \times 3 \times 7$. You can see it's possible to regroup those primes into 9×14, which subtract to 5. Those are the numbers you want, and here is what you're going to do with them. Set up the two binomials, beginning both with $6x$. Yes, you're absolutely right. They can't both be $6x$ or the trinomials would begin with $36x^2$, but that will fix itself in a moment.

$$\left(6x \quad\right)\left(6x \quad\right)$$

Now take the 14 and the 9, and think for a minute about signs. You want $+5x$ so use $+14$ and -9.

$$\left(6x + 14\right)\left(6x - 9\right)$$

Now it's time to get rid of that extra 6. Notice that there's a common factor of 2 in $6x + 14$ and a common factor of 3 in $6x - 9$. Factor those out, because they make the extra factor of 6. Throw them away and you'll have the correct factors.

$$2\left(3x + 7\right) \cdot 3\left(2x - 3\right)$$
$$\left(3x + 7\right)\left(2x - 3\right)$$

Factor each polynomial.

1. $3x^2 + 11x + 10$

2. $2x^2 - 3x + 1$

3. $2x^2 + 7x + 3$

4. $12x^2 + 32x + 5$

5. $6x^2 + 17x + 12$

6. $10x^2 - 49x - 5$

7. $9x^2 - 27x + 20$

8. $18x^2 + 15x + 2$

9. $15x^2 - 13x - 6$

10. $4x^2 - 29x + 30$

Special factoring patterns

Most of the time, factoring is a trial-and-error process, but there are a few cases where the problem is unusual and memorizing is a much better tactic. There are two factoring patterns that should be memorized.

Difference of squares

Most of the expressions you're asked to factor are trinomials, so you might not expect a binomial like $x^2 - 4$ or $9t^2 - 16$ to be factorable. It turns out that when you multiply the sum and difference of the same two terms with the FOIL rule, however, the outer and inner terms add to 0 and you produce a square minus a square. The difference of squares, $a^2 - b^2$, factors to $(a+b)(a-b)$, so $x^2 - 4 = (x+2)(x-2)$ and $9t^2 - 16 = (3t+4)(3t-4)$.

Perfect square trinomial

The perfect square trinomial is one you could figure out how to factor without any memorizing, but it's convenient to have it memorized to save time. When you square a binomial like $2x + 5$, the first and last terms are squares and the inner and outer are identical.

$$(2x+5)^2 = (2x+5)(2x+5)$$
$$= 2x \cdot 2x + 5 \cdot 2x + 5 \cdot 2x + 5 \cdot 5$$
$$= 4x^2 + 10x + 10x + 25$$
$$= \underbrace{4x^2}_{\text{square}} + \underbrace{20x}_{2 \cdot 5 \cdot 2x} + \underbrace{25}_{\text{square}}$$

So $(ax+b)^2 = (ax)^2 + 2abx + b^2$. When you see that the first and last terms of a trinomial are perfect squares, check the middle term to see if you have a perfect square trinomial.

Factor each polynomial.

1. $x^2 - 49$

2. $x^2 + 6x + 9$

3. $36t^2 - 1$

4. $9t^2 - 24t + 16$

5. $16 - y^2$

6. $9y^2 + 42y + 49$

7. $4x^2 - 81$

8. $4x^2 + 4x + 1$

9. $16a^2 - 9y^2$

10. $x^2 - 12x + 36$

Factoring by grouping

When you're asked to factor a binomial, you should look for a common factor first, and then, whether you factored out a common factor or not, ask yourself if the binomial is a difference of squares. There is nothing else (that you know about) that works on binomials. The common experience will be factoring trinomials. There again you should look for a GCF, and then try to factor to two binomials.

Occasionally, you'll be asked to factor a polynomial with more than three terms, and of course, you'll look for a common factor. There is one other tactic, called factoring by grouping, which may let you do further factoring, especially on polynomials with four terms.

The good news is that factoring by grouping just uses the factoring techniques you already know in a slightly different way. Take a look at $2x^3 - 5x^2 - 8x + 20$. That coefficient of −5 in the squared term means there's no factor common to all four terms, but don't give up. Break the polynomial into two groups of two terms each.

$$2x^3 - 5x^2 \mid -8x + 20$$

Factor the greatest common factor from the first two terms, and then factor out the greatest common factor from the second two terms.

$$x^2(2x - 5) \mid -4(2x - 5)$$

If, after you remove the common factors, the two binomials in the parentheses match, you can treat that binomial as a common factor. Think $y = 2x - 5$ and the expression is $x^2y - 4y = y(x^2 - 4)$. Put the $2x - 5$ back in there and you have $(2x - 5)(x^2 - 4)$. Notice that you can factor $x^2 - 4$. That won't always happen, but you should look for it.

$$2x^3 - 5x^2 - 8x + 20 = x^2(2x - 5) - 4(2x - 5)$$
$$= (2x - 5)(x^2 - 4)$$
$$= (2x - 5)(x + 2)(x - 2)$$

Use factoring by grouping to factor each polynomial as completely as possible.

1. $x^3 + 2x^2 - x - 2$

2. $x^3 - 3x^2 - 4x + 12$

3. $x^3 + 2x^2 - 100x - 200$

4. $2x^3 - x^2 + 8x - 4$

5. $3xy + 5y - 9x - 15$

6. $2xz + 3z + 4x + 6$

7. $5xy + 3y + 5x + 3$

8. $x^2y^2 - y^2 - 4x^2 + 4$

9. $x^2y^2 - 9x^2 + y^2 - 9$

10. $x^2y^2 + y^2 + 4x^2 + 4$

Calculator notes #6

The other operation you're often called up to perform when working with polynomials is factoring. You can check your factors by comparing your answer with the original problem as described in Calculator Notes #5, but what if you haven't gotten an answer because you're stumped?

First, make sure you've done a thorough, well-organized attempt to find factors on your own. If you're truly stuck, you'll want to know whether the polynomial has any real factors and if it does, whether they're rational, because you're not expected to guess at irrational numbers. Use an idea from Calculator Notes #1, where you found the x-intercept of the graph to find the solution of the equation. If you can find the solutions of a polynomial equation that way, you can work backwards to the factors.

- Go to the [Y=] menu and clear old equations.
- Enter the polynomial that you're trying to factor in the Y1= slot, and graph. Adjust the window so that you can see the x-intercepts. Expect as many x-intercepts as the degree of the polynomial, but there may be fewer.

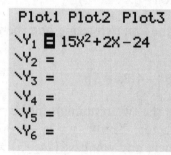
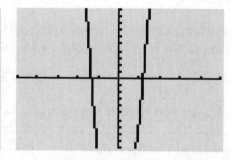

- If you see no x-intercepts, make sure it's not just a window problem. [ZOOM] 2: Zoom In and [ZOOM] 3: Zoom Out can help. If there really are no x-intercepts, this polynomial has no real zeros. There is no real number that you can plug in for x to make the polynomial equal 0. The polynomial cannot be factored.
- If you see one or more x-intercepts, use [2nd] [TRACE] 2: Zero to find them.
  - If the x-intercept is an integer, like 7, then your factor is $x - 7$.
  - If the x-intercept is not an integer, it will display as a decimal. If the decimal terminates, like 3.875, it's rational. If the decimal repeats a pattern, like 2.666666..., it's rational. If the x-intercept is an irrational number, its decimal does not terminate and

does not repeat any pattern, but the calculator will display as much of that decimal as it has space to show. You have to make the judgment call about whether you think the decimal you're looking at is rational or irrational.

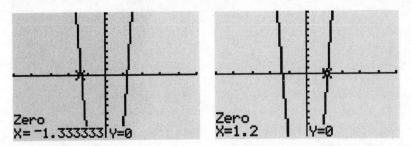

If the x-intercept is rational, you'll want to change it from decimal form to fraction form. Hopefully, you recognize the decimal equivalents of common fractions, but if not, type the decimal on the main screen of your calculator, and press $\boxed{MATH}$. Your cursor will be on 1: Frac. Press $\boxed{ENTER}$ twice and the fraction equivalent will display.

$$-1.333333333\triangleright Fr.\triangleright$$
$$-\frac{4}{3}$$

Once the x-intercept is in fraction form, start with (x − fraction), and then change to a more traditional format. If the x-intercept is $-\frac{4}{3}$, for example, start with $\left(x-\left(-\frac{4}{3}\right)\right)$ and simplify to $\left(x+\frac{4}{3}\right)$. Get to the more traditional form by multiplying both terms by the denominator: $\left(x+\frac{4}{3}\right) \Rightarrow (3x+4)$. If the x-intercept is 1.2, $(x-1.2) = \left(x-\frac{6}{5}\right) \Rightarrow (5x-6)$.

Radicals

8

Once you learn to perform a mathematical operation, it is common to think about how to reverse it. You undo addition with subtraction. You reverse multiplication by division. So once you learn about exponents, or raising a number to a power, it is natural to consider how to reverse that process. If you know the result of exponentiation, of raising a number to a power, how do you find your way back to the number that was the base of the power?

Powers and Roots

For every exponent, every power to which you can raise a base, there is a corresponding root to take you back to the original base. If, for some numbers a and b, $a^2 = b$, then a is the *square root* of b. Seven is the square root of 49 because $7^2 = 49$. The square root is the root you encounter most often, just as the second power, or square, is the exponent you see most often, but there is a third, or cube root, a fourth root, and a root to "undo" every power. Square roots are written using a *radical sign*: $\sqrt{49} = 7$ or, in general, $\sqrt{b} = a$. The expression under the radical sign is called the *radicand*. The word "radical" comes from the Latin word *radix*, meaning root.

You can write other roots in a similar fashion using other powers. Roots other than the square root are indicated by placing a small number, called the *index*, in the crook of the radical sign. For example, because 2^3 is equal to 8, two is the *cube root* of eight, written: $\sqrt[3]{8} = 2$. When no index is shown, the square root is assumed.

When the index of the radical is even, as in square roots, there are both positive and negative roots. Seven is the square root of 49 because $7^2 = 49$, but $(-7)^2 = 49$ as well, so 49 has two square roots, 7 and −7. The positive square root is considered the *principal square root*, and we agree that $\sqrt{b}$ will denote the principal root of b. We'll write $-\sqrt{b}$ if we want the negative square root or $\pm\sqrt{b}$ if we want both.

There is one only real root when the index is odd. The third, or cube, root of 64 is 4, because $4^3 = 64$ but $(-4)^3 = -64$ not 64. The cube root of −64 is −4.

Notice that, although you've probably learned there is no real square root of a negative number (and that's true for all even roots), there is a real cube root of a negative number (and that extends to all odd roots). Because $(-4)^3 = -64$, the cube root of −64 is −4.

	Positive number	Negative number
Even Root	2 real roots, 1 positive and 1 negative	No real roots
Odd Root	1 real positive root	1 real negative root

EXERCISE
8·1

Find each of the following roots, if possible. (Try not to use a calculator.)

1. $\sqrt{36}$

2. $-\sqrt{81}$

3. $\pm\sqrt{25}$

4. $\sqrt[3]{27}$

5. $\sqrt[3]{-125}$

6. $\sqrt{144}$

7. $-\sqrt[4]{16}$

8. $\sqrt[5]{-32}$

9. $\sqrt[3]{1,000}$

10. $\sqrt[4]{-81}$

Estimating roots

In the previous exercise, all of the radicands, the numbers under the radical sign, were chosen because they were perfect squares, or perfect cubes, or perfect whatever-power-matched-the-index. There are many, many numbers that don't fit that description. The numbers 81 and 100 are both perfect squares, 9^2 and 10^2, respectively, but that means that all the numbers between them are not. That means that many numbers will not have tidy integer square roots. (Although we're looking at square roots right now, the same can be said for other roots.)

Calculators are a tremendous help when working with roots, because they will give you an estimate of the root accurate to 7 or 8 decimal places. (Before calculators, you had to go look them up in a book of tables, and only get 3 or 4 decimal places.) But for some work, you don't need that much information.

Suppose you are working with $\sqrt{67}$. You know that $8^2 = 64$ and $9^2 = 81$, and you know 67 is between 64 and 81, closer to 64. That means $\sqrt{67}$ is between 8 and 9, closer to 8, and that may be enough to let you estimate your answer. You can do the same with other roots. If you know $3^3 = 27$ and $4^3 = 64$, then $\sqrt[3]{27} < \sqrt[3]{59} < \sqrt[3]{64}$. You can estimate that $\sqrt[3]{59}$ is between 3 and 4, and closer to 4.

Rational or irrational?

The Real Numbers is the name we give to the set of all numbers used in arithmetic. The Real Numbers include integers, fractions, and decimals, and you've probably seen that those subsets overlap. Integers can be written as fractions by giving them a denominator of 1, and fractions can be written as decimals. Integers, fractions, and decimals are subsets of the Real Numbers that overlap or intersect. The Real Numbers can be divided into two disjoint, or non-overlapping subsets, called the Rational Numbers and the Irrational Numbers.

The Rational Numbers are all numbers that can be written as a ratio of two integers. In simple language, they're numbers that can be written as fractions. The Irrational Numbers are the numbers that can't be written that way. Rational numbers are easy to think of : ½, −7, and 1.7 are examples. The number ½ is already a fraction, −7 can be $\frac{-7}{1}$ and 1.7 can be $\frac{17}{10}$. Not all decimals are rational, however. Decimals that terminate, like 0.5 or 1.7, can be written as fractions, and repeating decimals, like 0.3333… , which is equal to ⅓, are rational, but decimals that do not terminate and do not repeat cannot be expressed as the ratio of two integers and so are irrational.

Irrational numbers probably don't come to mind as quickly, so you may be surprised to know that there are more irrational numbers than rational numbers. Once you find one irrational number, that number times any rational number gives you another irrational number. So if you found 2 irrational numbers, you actually would have twice as many irrationals as rationals, and there are many more than 2 irrational numbers. Many roots are irrational, and therefore are decimals that don't terminate and don't repeat. For example, $\sqrt{2} \approx 1.414213562…$. That decimal goes on forever, and no matter how many digits of it you write, when you square that number, it won't give you exactly 2. It will come extremely close, but not exactly equal.

Exact or approximate?

Because you can't write all the digits of a decimal that doesn't end and doesn't repeat, when a root is an irrational number and you try to write it as a decimal, you only get an approximate answer. There are times when an approximate answer is good enough, but at other times, you'll want to express the exact value of the root. In those cases, you'll have to leave the number in radical form, but you'll want to try to rewrite it as a rational number times a simpler radical.

EXERCISE 8·2

Place each of the following roots between two consecutive integers. Then make your best guess at the approximate value of the root, to the nearest tenth. (Don't use your calculator to find the root. Make an educated guess. If you must, you can use your calculator to check.)

1. $\sqrt{11}$
2. $\sqrt{90}$
3. $\sqrt{150}$
4. $\sqrt{48}$
5. $\sqrt{3}$

6. $\sqrt{20}$
7. $\sqrt[3]{10}$
8. $\sqrt[3]{16}$
9. $\sqrt[3]{30}$
10. $\sqrt[3]{-81}$

Simplifying radical expressions

To make radicals easier to work with, you'll want to put them in simplest radical form. *Simplest radical form* means the expression contains only one radical, the radicand is as small as possible, and there is no radical in the denominator (or divisor) of a quotient. If there is any arithmetic to be done, do it, so that only one radical remains. The radical $\sqrt{48}$ is a single radical, but $48 = 16 \times 3$, and 16 is a perfect square, so it's possible to write $\sqrt{48}$ as $\sqrt{16 \cdot 3}$, which is equivalent to $4\sqrt{3}$. Always try to have the smallest possible radicand.

Simplest radical form

The principal rule for simplifying radicals tells us that $\sqrt{ab} = \sqrt{a}\sqrt{b}$. This rule lets you turn the product of two radicals into a single radical, and it lets you rewrite a radical as a multiple of a smaller radical. You can rewrite $\sqrt{18}\sqrt{2}$ as $\sqrt{36}$, or simply 6, and you can take $\sqrt{8}$ and realize that it's equivalent to $\sqrt{4}\sqrt{2}$, which is equal to $2\sqrt{2}$ because the square root of 4 is 2. By looking for perfect square factors of the radicand and applying this rule, you can rewrite radical expressions with smaller radicands.

When variables appear under radicals, it's still possible to simplify if you work carefully: $\sqrt[3]{x^3} = x$, and $\sqrt[3]{x^4} = \sqrt[3]{x^3 \cdot x} = \sqrt[3]{x^3} \cdot \sqrt[3]{x} = x\sqrt[3]{x}$. For even roots, like square roots, you need a bit more care. It's easy to think that $\sqrt{x^2} = x$ and you'll often see that written, but it's not technically correct. If we write $\sqrt{4^2}$, that's $\sqrt{16}$, which is 4. We said that $\sqrt{16}$ would mean the principal square root, so we only need 4, not −4. If we write $\sqrt{(-4)^2}$, however, that's also $\sqrt{16}$, or 4. Notice the original had −4, but we ended with 4. When you write $\sqrt{x^2}$, you don't know if x is a positive number or a negative number, but you want a positive result. To be sure we get that, we say $\sqrt{x^2} = |x|$, the absolute value of x.

EXERCISE
8·3

Put each expression in simplest radical form.

1. $\sqrt{32}$

2. $\sqrt{72}$

3. $\sqrt{a^3}$

4. $\sqrt{98}$

5. $\sqrt{8x^2}$

6. $\sqrt{12y^3}$

7. $\sqrt{50ab^5}$

8. $\sqrt{27x^3y^2}$

9. $\sqrt{49a^3}$

10. $\sqrt{48a^7b^3c^2}$

Rationalizing denominators

Because radicals are often irrational, having a radical in the denominator of a fraction can make it hard to estimate the quotient. If you have $\dfrac{\sqrt{5}}{2}$, you can think $\sqrt{5}$ is a little more than 2, so if you divide it by 2, you get a little more than 1. When you look at $\dfrac{2}{\sqrt{5}}$, on the other hand, it's harder to do that kind of approximation. That's one reason you don't want radicals in the denominator.

If the denominator of an expression is a radical or a multiple of a radical, you can remove the radical, or *rationalize the denominator*, by multiplying the numerator and denominator of the fraction by the radical in the denominator. This is the same method you use to express fractions with a common denominator. Multiplying both numerator and denominator by the same number is equivalent to multiplying by 1, so it changes the appearance of the fraction but not the value.

$$\frac{6}{\sqrt{5}} = \frac{6}{\sqrt{5}} \cdot \frac{\sqrt{5}}{\sqrt{5}} = \frac{6\sqrt{5}}{\sqrt{25}} = \frac{6\sqrt{5}}{5}$$

If the denominator is a sum or difference that includes a radical, you will still want to multiply the numerator and denominator by the same number, but multiplying by just the radical will not be effective. You will need to multiply by the *conjugate* of the denominator, the same two terms connected by the opposite sign. The conjugate of $3+\sqrt{2}$ is $3-\sqrt{2}$, and the conjugate of $\sqrt{7}-4$ is $\sqrt{7}+4$. Multiplying by the conjugate eliminates the radical because when you multiply the sum and difference of the same two terms using the FOIL rule, the middle terms, where the radicals would have been, add to 0.

$$\frac{4}{2+\sqrt{3}}=\frac{4}{2+\sqrt{3}}\cdot\frac{2-\sqrt{3}}{2-\sqrt{3}}=\frac{4\left(2-\sqrt{3}\right)}{\left(2+\sqrt{3}\right)\left(2-\sqrt{3}\right)}=\frac{8-4\sqrt{3}}{4-3}=8-4\sqrt{3}$$

EXERCISE
8·4

Rationalize the denominators and put each expression in simplest radical form.

1. $\dfrac{3}{\sqrt{6}}$

2. $\dfrac{24}{\sqrt{5}}$

3. $\dfrac{\sqrt{10}}{\sqrt{2}}$

4. $\dfrac{\sqrt{15}}{\sqrt{5}}$

5. $\dfrac{20\sqrt{20}}{4\sqrt{5}}$

6. $\dfrac{18}{\sqrt{5}-3}$

7. $\dfrac{20}{1-\sqrt{5}}$

8. $\dfrac{\sqrt{7}-5}{\sqrt{7}+5}$

9. $\dfrac{3}{\sqrt{5}-\sqrt{3}}$

10. $\dfrac{\sqrt{3}+4}{\sqrt{3}-2}$

Adding and subtracting radicals

Only like radicals can be combined by addition or subtraction, and they combine like variable terms, by adding or subtracting the coefficients, the numbers in front. $5\sqrt{2}+3\sqrt{2}=8\sqrt{2}$, but $\sqrt{2}+\sqrt{3}$ cannot be combined.

Problems that look like unlike radicals at first glance may simplify down to like radicals. When you need to combine radicals by addition or subtraction, first put each term in simplest radical form and then combine like radicals.

$$\sqrt{27}+\sqrt{48}=\sqrt{9}\sqrt{3}+\sqrt{16}\sqrt{3}=3\sqrt{3}+4\sqrt{3}=7\sqrt{3}$$

Simplify each expression as completely as possible.

1. $3\sqrt{3} + \sqrt{27}$

2. $\sqrt{5} + \sqrt{20}$

3. $\sqrt{48} - \sqrt{12}$

4. $\sqrt{200} + \sqrt{8}$

5. $\sqrt{45} + \sqrt{80}$

6. $2\sqrt{7} + \sqrt{28}$

7. $\sqrt{99} - 3\sqrt{44}$

8. $3\sqrt{96} - \sqrt{24}$

9. $\sqrt{27} + 5\sqrt{3} - \sqrt{48}$

10. $3\sqrt{45} - 2\sqrt{20} + \sqrt{125}$

Solving radical equations

Radical equations are equations that contain one or more radicals with the variable in the radicand. The key to solving radical equations is isolating the radical and then raising both sides of the equation to a power so that the radical sign is lifted.

One radical

To solve an equation containing one radical, isolate the radical by moving all terms that do not involve the radical to the other side. Square both sides of the equation, and solve the resulting equation. Be sure to check solutions in the original equation to eliminate any extraneous solutions. Extraneous solutions are values that do not satisfy the equation, even though you've done nothing wrong. They often turn up when solving radical equations.

$$5\sqrt{3x-2} + 4 = 29$$
$$5\sqrt{3x-2} = 25$$
$$\sqrt{3x-2} = 5$$
$$3x - 2 = 25$$
$$3x = 27$$
$$x = 9$$

Solve each equation. Check your answers in the original equations. If the equation cannot be solved, write No solution.

1. $\sqrt{x} = 3$

2. $\sqrt{y} = -4$

3. $\sqrt{x} - 9 = 0$

4. $-3 = \sqrt{3x}$

5. $3 = \sqrt{2y} - 7$

6. $3\sqrt{z} = -2$

7. $\sqrt{x-1} = 4$

8. $\sqrt{2x-3} - 7 = 0$

9. $2\sqrt{x+5} - 8 = 0$

10. $3\sqrt{4x-3} - 2 = 3$

Two radicals

When an equation contains two radicals, choose one to eliminate first. Isolate that radical, and square both sides. This may mean that you need to FOIL one side, and you'll probably find that while squaring the isolated radical eliminates that radical, FOIL multiplication on the other side will lift one radical sign but introduce another. Just isolate the remaining radical, square both sides again, and solve the resulting equation. Be sure to check solutions in the original equation. Extraneous solutions are common.

$$\sqrt{x+3} + \sqrt{x+10} = 7$$
$$\sqrt{x+3} = 7 - \sqrt{x+10}$$
$$\left(\sqrt{x+3}\right)^2 = \left(7 - \sqrt{x+10}\right)^2$$
$$x+3 = 49 - 14\sqrt{x+10} + x + 10$$
$$x+3 = x + 59 - 14\sqrt{x+10}$$
$$-56 = -14\sqrt{x+10}$$
$$4 = \sqrt{x+10}$$
$$16 = x + 10$$
$$6 = x$$

Solve each equation. Check for extraneous solutions.

1. $\sqrt{4x+5} = \sqrt{6x-5}$

2. $\sqrt{3x+1} - \sqrt{x+11} = 0$

3. $4\sqrt{2x+1} = 3\sqrt{3x+4}$

4. $1 + \sqrt{x^2-1} = x + 3$

5. $\sqrt{x+1} - 3 = -\sqrt{x-2}$

6. $\sqrt{2x-7} + 1 = \sqrt{2x+3}$

7. $\sqrt{x} - \sqrt{x+7} = -1$

8. $\sqrt{x-16} + 2 = \sqrt{x}$

9. $\sqrt{x+2} = \sqrt{x+4} - 2$

10. $\sqrt{x+5} = \sqrt{x-2} + 1$

Graphing square root equations

The graph of a square root equation looks like half of a parabola lying on its side. The graph of $y = \sqrt{x}$ begins at the origin and forms a slowly rising curve in the first quadrant as shown in Figure 8.1. A negative multiplier in front of the radical will flip the graph over the x-axis.

Figure 8.1 The top half of a parabola on its side is the graph of the square root function.

Because negative numbers have no square roots in the real numbers, the graph only exists for values of x that make the radicand positive. The graph of $y = \sqrt{x-h}$ begins at $(h, 0)$, and the graph of $y = \sqrt{x+h}$ begins at $(-h, 0)$. A constant added on the end of the equation moves the graph up or down. Consider where the graph begins before choosing x-values for your table of values.

EXERCISE
8·8

Graph each function by making a table of values and plotting points.

1. $y = \sqrt{x}$

2. $y = \sqrt{x-3}$

3. $y = \sqrt{x+1}$

4. $y = \sqrt{x-4}$

5. $y = \sqrt{x}+2$

6. $y = 2\sqrt{x}$

7. $y = -\sqrt{x}$

8. $y = 2\sqrt{x-3}+1$

9. $y = 4-\sqrt{x+2}$

10. $y = -3\sqrt{2x+5}-4$

Quadratic equations and their graphs

·9·

Quadratic equations are equations that contain a term in which the variable is squared. The standard form of a quadratic equation is $ax^2 + bx + c = 0$, but you may have to do some rearranging to get the equation you're given into that form. The x-term or the constant term may be missing if b equals 0 or c equals 0, but if the x^2 term is missing, it's not a quadratic equation. The graph of a quadratic equation $y = ax^2 + bx + c$ has a particular shape called a parabola.

Solving by square roots

If a quadratic equation contains just a squared term and a constant term, you can solve it by moving the terms to opposite sides of the equal sign and taking the square root of both sides. Remember that there is both a positive and a negative square root of any positive number.

$$2x^2 - 64 = 0$$
$$2x^2 = 64$$
$$x^2 = 32$$
$$x = \pm\sqrt{32}$$
$$= \pm 4\sqrt{2}$$

If the constant is not a perfect square, leave solutions in simplest radical form, unless there's a very good reason to use a decimal approximation.

EXERCISE
9·1

Solve each equation by isolating the square and taking the square root of both sides.

1. $x^2 = 64$

2. $x^2 - 16 = 0$

3. $x^2 - 8 = 17$

4. $x^2 = 18$

5. $3x^2 = 48$

6. $t^2 - 1000 = 0$

7. $2y^2 - 150 = 0$

8. $9x^2 = 4$

9. $64y^2 = 25$

10. $4x^2 - 15 = 93$

Completing the square

If the quadratic equation has just a square term and a constant term, you can solve it by taking the square root of both sides. If it has three terms that happen to form a perfect square trinomial, you can rewrite it as the square of a binomial and then solve by taking the square root of both sides. Much of the time, however, the polynomial is not a perfect square.

Completing the square is a process that turns one side of the equation into a perfect square trinomial so that you can solve by taking the square root of both sides. Of course, completing the square doesn't just magically change one side of the equation. It adds the same number to both sides so that the new equation is equivalent to the original. The key is to know what to add.

To complete the square, move the constant to the opposite side of the equation from the x^2 and x terms.

$$3x^2 + 24x + 15 = 0$$
$$3x^2 + 24x = -15$$

Divide both sides of the equation by the coefficient of x^2.

$$3x^2 + 24x = -15$$
$$x^2 + 8x = -5$$

Take half the coefficient of x, square it, and add the result to both sides. Write the polynomial as a square.

$$x^2 + 8x = -5$$
$$x^2 + 8x + 4^2 = -5 + 4^2$$
$$x^2 + 8x + 16 = -5 + 16$$
$$(x+4)^2 = 11$$

Solve the equation by taking the square root of both sides and then isolating the variable.

$$(x+4)^2 = 11$$
$$x + 4 = \pm\sqrt{11}$$
$$x = -4 \pm \sqrt{11}$$

**EXERCISE
9·2**

Solve each equation. Do not FOIL the binomial square. Take the square root of both sides. Give answers in simplest radical form.

1. $(x-2)^2 = 25$
2. $(x+1)^2 = 9$
3. $(x-3)^2 = 48$
4. $(x+1)^2 = 75$
5. $(3x-5)^2 = 12$

Complete the square and solve the equation.

6. $y^2 - 8y - 7 = 0$
7. $x^2 - 5x = 14$
8. $x^2 + 4x - 4 = 0$
9. $a^2 + 5a - 3 = 0$
10. $t^2 = 10t - 8$

The quadratic formula

Completing the square is a very effective method for solving quadratic equations, but it can get complicated and the numbers can get messy, so you soon find yourself wishing for an easier way. The quadratic formula is a shortcut to the solution you would have obtained by completing the square. Using it is a bit complicated, but easier than doing all the work of completing the square.

If $ax^2 + bx + c = 0$, then $x = \dfrac{-b \pm \sqrt{b^2 - 4ac}}{2a}$. You just pick the values of a, b, and c out of the equation, plug them into the formula, and simplify. Be certain your equation is in $ax^2 + bx + c = 0$ form before deciding on the values of a, b, and c.

To solve $5x = 2 - 3x^2$, first put the equation in standard form. Compare $3x^2 + 5x - 2 = 0$ to $ax^2 + bx + c = 0$, and you find that $a = 3$, $b = 5$, and $c = -2$. Plug those values into the formula.

$$x = \frac{-b \pm \sqrt{b^2 - 4ac}}{2a}$$

$$= \frac{-5 \pm \sqrt{5^2 - 4 \cdot 3 \cdot -2}}{2 \cdot 3}$$

Follow the order of operations and watch your signs as you simplify.

$$x = \frac{-5 \pm \sqrt{25 + 24}}{6}$$

$$= \frac{-5 \pm \sqrt{49}}{6}$$

$$= \frac{-5 \pm 7}{6}$$

The two solutions that are typical of quadratic equations come from that $\pm$ sign.

$$x = \frac{-5 \pm 7}{6}$$

$$x = \frac{-5 + 7}{6} \qquad x = \frac{-5 - 7}{6}$$

$$= \frac{2}{6} \qquad \text{or} \qquad = \frac{-12}{6}$$

$$= \frac{1}{3} \qquad\qquad\qquad = -2$$

The discriminant

One portion of the quadratic formula, called the discriminant, can give you useful information about the number and type of solutions your equation has. The radicand, $b^2 - 4ac$, tells you whether you have one real solution, two real solutions, or no real solutions, and sometimes that's all you need to know.

If $b^2 - 4ac$ is positive, the equation has two real solutions. You have a positive number under the radical, so you get the positive and the negative square root and produce two solutions.

If $b^2 - 4ac$ equals 0, the positive and negative square roots are both 0, so you end up with only one solution, $x = \dfrac{-b}{2a}$. This is sometimes called a double root, since it really is two solutions that are exactly the same. If $b^2 - 4ac$ is negative, you know it's impossible to find the square root of a negative number in the real numbers. (Later in your math career, you'll learn about a set of numbers larger than the reals, where negative numbers do have square roots.) So if the discriminant is negative, there are no real solutions.

You can get one more bit of information from the discriminant. If the discriminant is positive and it's a perfect square, the two solutions will be rational numbers, but if the discriminant is positive and not a perfect square, the two solutions will be irrational.

EXERCISE 9·3

Solve each equation by the quadratic formula. If necessary, leave answers in simplest radical form.

1. $x^2 + 4x - 21 = 0$

6. $t^2 + 6t - 15 = 0$

2. $t^2 = 10 - 3t$

7. $4x^2 - 3 = x$

3. $y^2 - 4y = 32$

8. $3x^2 - 1 = 2x$

4. $x^2 = 6 + x$

9. $x + 5 = 3x^2 - x$

5. $6x + x^2 = 9$

10. $6x^2 - 2 = x$

Use the discriminant to tell whether each equation has two rational solutions, two irrational solutions, one rational solution, or no real solutions.

11. $x^2 + 5x - 9 = 0$

16. $2x^2 + 5x + 3 = 0$

12. $x^2 - 3x + 5 = 0$

17. $5x^2 + 12x + 5 = 0$

13. $x^2 + 3x - 4 = 0$

18. $2x^2 + 3x + 5 = 0$

14. $x^2 + 6x + 9 = 0$

19. $4x^2 + 4x + 1 = 0$

15. $3x^2 + 6x + 1 = 0$

20. $-3x^2 + 3x + 5 = 0$

Solving by factoring

The quadratic formula will give you solutions for any quadratic equation, but it may require complicated calculations. Sometimes there's no way around that. Completing the square is complicated, the quadratic formula can be complicated, and irrational solutions aren't easy. In other cases, however, the quadratic polynomial can be factored and the equation can be solved by applying the zero product property. If the product of two factors is 0, then at least one of the factors is 0. If the quadratic can be factored, that's probably the easiest way to solve.

First, put the equation in standard form. With all terms on one side of the equation equal to 0, factor the quadratic expression. Set each factor equal to 0 and solve the resulting equations.

$$5x = 4 - 6x^2$$
$$6x^2 + 5x - 4 = 0$$
$$(2x - 1)(3x + 4) = 0$$

$$2x - 1 = 0 \qquad\qquad 3x + 4 = 0$$
$$2x = 1 \qquad \text{or} \qquad 3x = -4$$
$$x = \frac{1}{2} \qquad\qquad x = \frac{-4}{3}$$

EXERCISE 9·4

Solve each equation by factoring.

1. $x^2 + 5x + 6 = 0$

2. $x^2 + 12 = 7x$

3. $y^2 + 3y = 8 + y$

4. $a^2 - 3a - 4 = 6$

5. $20 = x^2 + x$

6. $x^2 + 5 = 6x$

7. $x^2 + 3x = 0$

8. $x^2 = 5x$

9. $x^2 - 1 = 3x^2 - 3x$

10. $2x^2 - x - 15 = 0$

Graphing quadratic functions

Graphing an equation of the form $y = ax^2 + bx + c$ produces a cup-shaped graph, called a parabola. Some information about the graph can be gathered from the equation without much effort and can help you construct a table of values and plot the graph. If a is a positive number, the parabola opens upward; if a is negative, the parabola opens downward (see Figure 9.1). The x-intercepts are the solutions of the equation $ax^2 + bx + c = 0$ and the y-intercept is $(0, c)$.

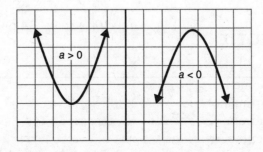

Figure 9.1 The coefficient of the square term tells you whether the parabola opens up or down.

Finding the axis of symmetry and vertex

The axis of symmetry of a parabola is an imaginary line through the center of the parabola. The parabola is symmetric about this line. If you were to fold the graph along the axis of symmetry, the two sides of the parabola would match. The vertex, or turning point, of the parabola sits right on the axis of symmetry. The equation of the vertical line that is the axis of symmetry is $x = \dfrac{-b}{2a}$. Once the axis of symmetry is known, plugging that x-value into the equation of the parabola will give you the y-coordinate of the vertex.

To graph $y = x^2 + 4x + 1$, notice that the x^2-term is positive, so the parabola will open up and its y-intercept will be (0, 1). You'll need the quadratic formula to find the x-intercepts.

$$
\begin{aligned}
x &= \frac{-b \pm \sqrt{b^2 - 4ac}}{2a} \\[2mm]
&= \frac{-4 \pm \sqrt{4^2 - 4 \cdot 1 \cdot 1}}{2 \cdot 1} \\[2mm]
&= \frac{-4 \pm \sqrt{16 - 4}}{2} = \frac{-4 \pm \sqrt{12}}{2} = \frac{-4 \pm 2\sqrt{3}}{2} = -2 \pm \sqrt{3}
\end{aligned}
$$

This is a good time to use a decimal approximation, since the x-axis is not usually marked with radicals. So $-2 + \sqrt{3} \approx -0.3$ and $-2 - \sqrt{3} \approx -3.7$. You'll have to estimate those x-intercepts, but even an estimate will help you line up the graph correctly.

The axis of symmetry for $y = x^2 + 4x + 1$ is $x = \dfrac{-b}{2a} = \dfrac{-4}{2 \cdot 1} = -2$, so the parabola will be symmetric across the vertical line $x = -2$. Plug -2 into $y = x^2 + 4x + 1$, and you find that the y-coordinate of the vertex is $y = (-2)^2 + 4(-2) + 1 = 4 - 8 + 1 = -3$. The vertex is $(-2, -3)$.

EXERCISE
9·5

Find the x- and y-intercepts of each parabola.

1. $y = x^2 - 4x + 3$

2. $y = x^2 - 4x - 5$

3. $y = x^2 + 2x$

4. $y = x^2 - 7x + 12$

5. $y = 2x^2 + x - 1$

Find the axis of symmetry and vertex of each parabola.

6. $y = x^2 - 8x + 15$

7. $y = x^2 + 4x - 2$

8. $y = 2x^2 - 4x + 3$

9. $y = -x^2 + 6x - 7$

10. $y = -x^2 + 4x + 7$

Table of values

Building a table of values is the most fundamental method of drawing the graph of an equation. Choosing values for x and substituting each one into the equation to find the corresponding value of y will let you plot enough points to graph any equation. When graphing a quadratic equation (or other nonlinear equation), it's important to make wise choices of x-values. Finding the axis of symmetry, vertex, and intercepts first will tell you where the interesting part of the graph is so that you can choose x-values on both sides of the vertex, but not too far away.

In the previous example, we were getting ready to graph $y = x^2 + 4x + 1$, and we already had four points on the graph. We knew the approximate values of the two x-intercepts, we knew the y-intercept, and we knew the vertex.

x	−3.7	−2	−0.3	0
y	0	−3	0	1

Choose a few more values for x on both sides of the vertex to finish your table. Plot the points and connect them with a smooth curve as shown in Figure 9.2.

x	−4	−3.7	−3	−2	−1	−0.3	0
y	1	0	−2	−3	−2	0	1

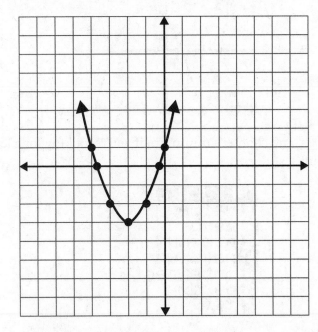

Figure 9.2 The graph of $y = x^2 + 4x + 1$.

EXERCISE
9·6

Graph each quadratic function. Use the vertex and intercepts to help make a table of values.

1. $y = 2x^2 - 1$

2. $y = -x^2 + 8x$

3. $y = x^2 + 2x - 15$

4. $y = x^2 - x - 2$

5. $y = x^2 - 4x + 3$

6. $y = -x^2 + 2x + 5$

7. $y = x^2 + 6x + 9$

8. $y = 4 - x^2$

9. $y = 2x^2 + x - 1$

10. $y = x^2 - 9$

Calculator notes #7

In Calculator Notes #1, we talked about finding the *x*-intercepts of graphs. When the graph you're looking at is a parabola, the graph of a quadratic equation, you may find one *x*-intercept, more than one *x*-intercept or no *x*-intercepts at all. The process of finding those *x*-intercepts is the same as for a linear equation. You just need to make sure that you place your cursor to the left, and then to the right, of the intercept you're trying to find. Think of it as fencing in the point you want. Then move on to another intercept and repeat, fencing in the second intercept point.

The other task that comes up with parabolas is finding the turning point, called the vertex. The vertex is either the lowest point on the graph, if the parabola opens up, or the highest point, if the parabola opens down.

To find the vertex, graph the equation and use the maximum or minimum function on your calculator:

- Move your cursor to the left of the vertex.

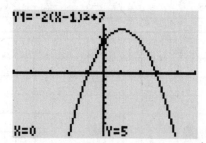

- Press [2nd] [TRACE] and choose 3: minimum if you're looking for a low point or 4: maximum for a high point.

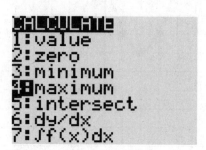

- Your cursor should already be left of the vertex, so press [ENTER].
- Move the cursor past the vertex and press [ENTER].

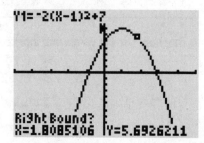

- If you wish, move your cursor closer to the vertex and press [ENTER].
- The coordinates of the vertex are displayed at the bottom of the screen.

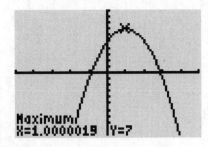

When a minimum or maximum value sits right on the x-axis, that x-intercept is not only the vertex; it's also a double zero. The quadratic equation represented by this parabola is a square and, rather than two different solutions, has the same solution twice. The zero function on your calculator may not be able to cope with this, because it expects the left and right boundaries to have opposite signs, and that doesn't happen with a double zero. You can try slowly tracing on the graph until you see the y-coordinate become zero, but that's tedious and you may never get exactly on zero. Instead, try this:

- With your original equation in Y1, also graph Y2 = 0. The graph will not look any different, because Y2 = 0 sits right on the x-axis.
- Press 2nd TRACE and choose 5:intersect. The two curves are Y1, the original equation, and Y2, the x-axis, and the intersection will be that point where the original equation touches the x-axis.

Proportion and variation ·10·

Mathematics is more than arithmetic, as you certainly have seen. One of the things math often does is compare quantities, and those comparisons can be based on different operations. The statement that x is 4 more than y is based on adding or subtracting, but saying a is 3 times the size of b is based on multiplication or division. A ratio is a comparison of two numbers by division. The relationship between 10 and 5 or between 26 and 13 can be expressed as a ratio: $10:5$ or $\frac{26}{13}$. Both of these are equal to $2:1$.

Using ratios and extended ratios

If two quantities are in the ratio $a:b$, it's not assured that they are exactly equal to a and b, but they are multiples of a and b. As a result, you can represent them as ax and bx and use those expressions to write an equation about the numbers. If two numbers add to 50 and are in ratio $3:7$, you can represent the numbers as $3x$ and $7x$ and write $3x + 7x = 50$. You'll find that $x = 5$, so the actual numbers are $3x = 3 \cdot 5$ and $7x = 7 \cdot 5$, or 15 and 35.

An extended ratio is a comparison of three or more numbers, usually written in the form $a:b:c$. If the measurements of the angles of a triangle are in ratio $2:3:5$, you can represent the measures of the angles by $2x$, $3x$, and $5x$ and add $2x + 3x + 5x = 180°$. Once you solve and find $x = 18°$, remember to multiply by the appropriate coefficients to find the angle measures: $2x = 36°$, $3x = 54°$, and $5x = 90°$.

EXERCISE 10·1

Use ratios to solve each problem.

1. Find the number of degrees in each angle of a triangle if the angles are in the ratio $3:4:5$.

2. A piece of wood 20 ft long needs to be cut into two pieces that are in ratio $2:3$. How long should each piece be?

3. Two numbers are in ratio $8:3$ and their difference is 65. Find the numbers.

4. Laura determined that the perfect recipe for her raspberry limeade was to mix raspberry juice and lime juice in a $5:7$ ratio. How much raspberry juice will she need to make 48 oz of the mixture?

5. Three numbers are in ratio 3 : 4 : 8. The sum of the two larger numbers exceeds twice the smallest by 48. Find the numbers.

6. Two numbers are in the ratio 5 : 6. If 8 is added to each of the numbers, they will be in the ratio 7 : 8. Find the numbers.

7. Two numbers are in the ratio 3 : 7. If 1 is added to the smaller number and 7 is added to the larger, they will be in the ratio 1 : 3. Find the numbers.

8. What should be added to both 9 and 29 to produce numbers that are in the ratio 3 : 4?

9. The numerator and denominator of a fraction are in the ratio 2 : 5. If 2 is subtracted from both the numerator and denominator, the resulting fraction is equal to $\frac{1}{3}$. Find the original numerator and denominator.

10. The larger of two numbers is 2 more than 3 times the smaller. If 3 is added to the smaller number and 1 is added to the larger, they will then be in the ratio 3 : 7. Find the numbers.

Solving proportions

A proportion is a statement that two ratios are equal, or an equation of the form $\frac{a}{b} = \frac{c}{d}$. Any two equal ratios form a proportion. In a proportion like 10 : 5 = 2 : 1, the numbers on the ends, 10 and 1, are called the extremes, and the numbers in the middle, 5 and 2, are called the means. When the ratios are written as fractions, the proportion is $\frac{\text{extreme}}{\text{mean}} = \frac{\text{mean}}{\text{extreme}}$.

In any proportion, the product of the means is equal to the product of the extremes. For example, in 10 : 5 = 2 : 1, $10 \times 1 = 5 \times 2$ and in $\frac{3}{7} = \frac{12}{28}$, $3 \times 28 = 7 \times 12 = 84$. When one term of the proportion is unknown, you can cross-multiply to create an equation that you can solve for the missing term of a proportion.

$$\frac{3}{x} = \frac{5}{35}$$

$$5x = 3 \cdot 35 = 105$$

$$x = \frac{105}{5} = 21$$

EXERCISE 10·2

Solve each proportion to find the value of the variable.

1. $\frac{5}{3} = \frac{7}{x}$

2. $\frac{w}{5} = \frac{6}{2}$

3. $\frac{9}{15} = \frac{15}{x}$

4. $\frac{18}{2.5} = \frac{30}{x}$

5. $\dfrac{4}{x+5} = \dfrac{3}{x}$

6. $\dfrac{x-7}{2} = \dfrac{2x-3}{5}$

7. $\dfrac{x}{4} = \dfrac{16}{x}$

8. $\dfrac{3}{x-4} = \dfrac{x+4}{3}$

9. $\dfrac{5}{x-12} = \dfrac{x+12}{5}$

10. $\dfrac{7}{x-3} = \dfrac{x}{4}$

Variation

Variation looks at how quantities change, specifically in relation to one another. When you observe two related variables, does one increase as the other increases? Do they increase at the same rate, or different rates? How do you describe that relationship? Or does one get larger as the other gets smaller? How do you describe that activity with an equation? There are two basic variation relationships, direct variation, when both increase (or both decrease), and inverse variation, when the quantities move in opposite directions. There are more complex relationships, of course, but those can be built by combining direct and inverse variation in different ways.

Direct variation

When two quantities vary directly, they increase or decrease together. If 2 hamburgers cost $3 and 4 hamburgers cost $6, the total cost of the hamburgers varies directly with the number of hamburgers you buy. The number of hamburgers goes up, and the total cost goes up. But the number of hamburgers increased by 2 hamburgers, and the total cost increased by $3. There has to be a number in the equation that talks about how much things change. If y varies directly as x, there is a constant k such that $y = kx$, or $\dfrac{y}{x} = k$. This constant of variation, as it's called, is the ratio of a y-value to its corresponding x-value. In the hamburger example, $\dfrac{\$3}{2 \text{ burgers}} = \dfrac{\$6}{4 \text{ burgers}}$ so $k = 1.5$. You can say total cost $= 1.5 \times$ number of burgers.

If you know that two quantities are directly related, you can plug in known values of x and y to find k, and once you know k, you can apply the relationship to other values of x or y. For example, if y varies directly as x, and $y = 12$ when $x = 2$, you can find that $k = 6$, either by solving $12 = k \cdot 2$ or by dividing $\dfrac{12}{2} = k$. Once you know that $k = 6$, if you're told that x has changed to 5, you can determine that $y = 6 \cdot 5 = 30$. If you find that y has changed to 42, you can solve $42 = 6x$ and see that $x = 7$.

EXERCISE 10·3

Use the direct variation equation y = kx *to find k, and then find the value of the variable requested.*

1. If y varies directly as x and $y = 12$ when $x = 4$, find y when $x = 14$.

2. If y varies directly as x and $y = 5$ when $x = 20$, find x when $y = 25$.

3. If t varies directly as r and $t = 52$ when $r = 13$, find t when $r = 78$.

4. If a varies directly as b and $a = 17$ when $b = 51$, find b when $a = 425$.

5. If y varies directly as the square of x and $y = 28$ when $x = 2$, find y when $x = 10$.

6. The voltage V in an electric circuit varies directly with the current I when $I = 40$ A, $V = 0.06$ V. Find V when $I = 6$ A.

7. The distance covered in a fixed time varies directly with the speed of travel. If you can travel 117 mi at 45 mph, how far will you travel in the same time if you increase your speed to 55 mph?

8. The time that passes between the moment a flash of lightning is seen and the moment a clap of thunder is heard varies directly with the observer's distance from the center of the storm. If 10 s elapse between the lightning and the thunder, the storm is 2 mi away. How far is the storm if 3 s pass between the flash and the sound of thunder?

9. The volume of a gas under a constant pressure varies directly with its temperature. At 18°C, a gas has a volume of 152 cm³. What is the volume when the temperature is 36°C?

10. If a car uses 7 gal of gas to travel 119 mi at a certain speed, how far can it travel on 10 gal of gas if it travels at the same speed?

Inverse variation

Quantities that vary inversely move in opposite directions. When one quantity increases, the other decreases. If y varies inversely as x, then there is a constant k such that $y = \dfrac{k}{x}$, or $xy = k$. You can use known values to find k and then calculate x or y, just as you did for direct variation, if you're given the value of the other variable. If y varies directly with x, and $y = 9$ when $x = 2$, substituting those values into $xy = k$ tells you that $k = 18$. If x increases to 6, $y = \dfrac{18}{6} = 3$.

EXERCISE

10·4

Use the inverse variation equation $y = \dfrac{k}{x}$ to find k, and then find the value of the variable requested.

1. If y varies inversely with x and $y = 4$ when $x = 6$, find y when $x = 8$.

2. If y varies inversely with x and $y = 8$ when $x = 4$, find x when $y = 2$.

3. If y varies inversely with x and $y = 24$ when $x = 3$, find y when $x = 9$.

4. If r varies inversely with t and $t = 11$ when $r = 12$, find t when $r = 3$.

5. If a varies inversely with b and $a = 54$ when $b = 10$, find a when $b = 45$.

6. Given that v varies inversely with t, and $v = 32$ when $t = 4$, what is the value of v when $t = 24$?

7. If w varies inversely with l and $w = 22$ when $l = 4$, what is the value of w when $l = 22$.

8. If N varies inversely with B and $N = 3$ when $B = 8$, find the value of B for which $N = 144$.

9. If y varies inversely with x and $y = 39$ when $x = 6$, find the value of x that produces a y-value of 18.

10. If $a = 18$ and $b = 16$, and a and b vary inversely, what is the value of a when b is 12?

Joint and combined variation

A quantity *varies jointly* with two or more other quantities if it varies directly with their product; that is, y varies jointly with x and z if there is a constant k such that $y = kxz$. If x is held constant, y varies directly with z. If z is held constant, y varies directly with x.

Combined variation occurs when y varies directly with x and inversely with z. The equation for combined variation has the form $y = \dfrac{kx}{z}$. If you hold x constant, y will vary inversely with z. If you hold z constant, y will vary directly with x.

In both joint and combined variation, you can find the value of k just as you did with direct and inverse variation. Choose the correct equation, plug in the known values, and solve for k. Then, just as before, if one or more variables change, plug in the new values along with k and solve for whatever is missing.

EXERCISE
10·5

Use the appropriate variation equation to find k, and then find the value of the variable requested.

1. If y varies jointly with x and z, and $y = 84$ when $x = 7$ and $z = 3$, find y when $x = 9$ and $z = 5$.

2. If y varies jointly with x and z, and $y = 165$ when $x = 6$ and $z = 11$, find x when $y = 200$ and $z = 10$.

3. If y varies jointly with x and z, and $y = 3375$ when $x = 3$ and $z = 9$, find z when $x = 12$ and $y = 10,500$.

4. If y varies directly with x and inversely with z, and $y = 9$ when $x = 270$ and $z = 12$, find y when $x = 100$ and $z = 8$.

5. If y varies directly with x and inversely with z, and $y = 53$ when $x = 3$ and $z = 12$, find x when $y = 159$ and $z = 28$.

6. If y varies directly with x and inversely with z, and $y = 4$ when $x = 8$ and $z = 14$, find z when $x = 36$ and $y = 6$.

7. The volume of a solid varies jointly with the area of its base and its altitude. When the area of the base is 12 cm² and the altitude is 16 cm, the volume is 64 cm³. Find the altitude when the volume is 60 cm³ and the area of the base is 9 cm².

8. The gravitational force between two bodies varies directly with the masses of the bodies and inversely with the square of the distance between them. When a 70-kg person stands on the surface of the earth, 6378 km from the center of the earth, the force of gravity is 686 N. Find the force of gravity acting on the same person when the person is in a plane 40,000 ft (about 12 km) above the surface of the earth.

9. The volume of a gas varies directly with temperature and inversely with pressure. At a temperature of 450°F and a pressure of 40 psi, a gas has a volume of 10 ft³. What is its volume if temperature is reduced to 440°F and pressure is raised to 50 psi?

10. The electric resistance of a wire varies directly with its length and inversely with the square of its diameter. A wire 80 ft long with a diameter of $\frac{1}{8}$ in. has a resistance of $\frac{1}{2}$ ohm. What is the resistance in a piece of the same type of wire that is 120 ft long and has a diameter of $\frac{1}{4}$ in.?

Rational equations and their graphs

A rational expression is the quotient of two polynomials, and a rational equation is an equation containing rational expressions. Since division by 0 is impossible, the denominator of a quotient can never be 0. The domain of a rational expression is the set of all values of the variable for which the expression is defined, which means the values of the variable that do not make the denominator 0.

Simplifying rational expressions

The quotient of two polynomials can look very complicated, but just like fractions, rational expressions can often be simplified. When we work with fractions, we often talk about reducing the fraction to lowest terms. That wording is a little bit of a misrepresentation, however. Reducing is making something smaller, and we don't change the value of the fraction at all. We just change the way it looks.

To simplify a rational expression, factor the numerator and the denominator and cancel any factors that appear in both. The rational expression $\dfrac{4x^2-8x-21}{8x^2+22x+15}$ looks like it would be difficult to work with, but if you pause for a minute and examine the numerator and denominator separately, you'll see that both factor.

$$\frac{4x^2-8x-21}{8x^2+22x+15}=\frac{(2x-7)(2x+3)}{(4x+5)(2x+3)}$$

Since both the numerator and the denominator have a factor of $2x+3$, you can think of $\dfrac{(2x-7)(2x+3)}{(4x+5)(2x+3)}$ as $\dfrac{2x-7}{4x+5}\cdot\dfrac{2x+3}{2x+3}=\dfrac{2x-7}{4x+5}\cdot 1$, or simply $\dfrac{2x-7}{4x+5}$.

EXERCISE

11·1

Reduce each expression to simplest form.

1. $\dfrac{14x+6}{7x+3}$

2. $\dfrac{3x-6}{9x-18}$

3. $\dfrac{a^2-49}{a^2-a-42}$

4. $\dfrac{2y^2-18y}{y^2-18y+81}$

5. $\dfrac{4y+32}{2y^2-128}$

6. $\dfrac{2x^2+4x-48}{2x^2-32}$

7. $\dfrac{x^2-4}{5x^2-11x+2}$

8. $\dfrac{x^2-16}{x^2+3x-4}$

9. $\dfrac{x^2+8x+15}{x^2+2x-3}$

10. $\dfrac{a^2+2a}{a^2-3a-10}$

Multiplying rational expressions

The basic rule for multiplying rational expressions is the same as the basic rule for fractions: numerator times numerator and denominator times denominator. Just as with fractions, however, much time and effort can be saved by "canceling," or simplifying, before multiplying.

To multiply rational expressions:

- Factor all numerators and denominators
- Cancel any factor that appears in both a numerator and a denominator
- Multiply numerator times numerator and denominator times denominator

Multiplying $\dfrac{4x^2-2x-30}{4x^2-5x+1}\cdot\dfrac{4x^2+35x-9}{4x^2-36}$ might look like an unreasonable task, but begin by focusing on numerators and denominators, one by one, and factoring as completely as possible.

$$\frac{4x^2-2x-30}{4x^2-5x+1}\cdot\frac{4x^2+35x-9}{4x^2-36}=\frac{2(2x+5)(x-3)}{(x-1)(4x-1)}\cdot\frac{(x+9)(4x-1)}{4(x+3)(x-3)}$$

Once you've factored, you can see that there are factors that appear more than once. Cancel them out, one from a numerator with one from a denominator.

$$\frac{\cancel{2}(2x+5)\cancel{(x-3)}}{(x-1)\cancel{(4x-1)}}\cdot\frac{(x+9)\cancel{(4x-1)}}{2\cancel{2}(x+3)\cancel{(x-3)}}$$

What's left will still require some work to multiply, but it's much easier than the original problem.

$$\frac{(2x+5)}{(x-1)}\cdot\frac{(x+9)}{2(x+3)}=\frac{2x^2+23x+45}{2x^2+4x-6}$$

EXERCISE
11·2

Multiply the rational expressions and give answers in simplest form.

1. $\dfrac{6x-24}{9x+27}\cdot\dfrac{3x+9}{2x-8}$

2. $\dfrac{x^2-25}{x^2-x-2}\cdot\dfrac{x^2+x-6}{x-5}$

3. $\dfrac{7x^2-7x}{4x^2}\cdot\dfrac{4x-12}{x^2-4x+3}$

4. $\dfrac{a^2+a-20}{a^2-16}\cdot\dfrac{3a^2-27}{a^2+a-6}$

5. $(25 - y^2) \cdot \dfrac{15 - 3y}{3y + 15}$

8. $\dfrac{6x^2 - x - 12}{4x^2 - 9} \cdot \dfrac{4x + 6}{3x + 4}$

6. $\dfrac{x^2 - 4}{3x^2 + x^3} \cdot \dfrac{x^3 - x^2 - 12x}{x^2 - 2x - 8}$

9. $\dfrac{x^2 - 25}{8x^2} \cdot \dfrac{4x}{x - 5}$

7. $\dfrac{x^2 - 6x + 5}{x - 5} \cdot \dfrac{x - 9}{x - 1}$

10. $\dfrac{12x^2}{12x - 18} \cdot \dfrac{4x^2 - 2x - 3}{4 + 8x}$

Dividing rational expressions

Take a minute to think about how you divide fractions. In fact, you don't. You multiply by the reciprocal of the divisor. Use the same tactic to divide rational expressions.

To divide rational expressions, invert the divisor and multiply. Factor all the numerators and denominators and cancel where possible.

$$\begin{aligned}
\frac{5x - 20}{x^2 + x - 12} \div \frac{x^2 - 2x - 8}{3x + 12} &= \frac{5x - 20}{x^2 + x - 12} \cdot \frac{3x + 12}{x^2 - 2x - 8} \\
&= \frac{5(x - 4)}{(x + 4)(x - 3)} \cdot \frac{3(x + 4)}{(x - 4)(x + 2)} \\
&= \frac{5\cancel{(x - 4)}}{\cancel{(x + 4)}(x - 3)} \cdot \frac{3\cancel{(x + 4)}}{\cancel{(x - 4)}(x + 2)} \\
&= \frac{5 \cdot 3}{(x - 3)(x + 2)} = \frac{15}{x^2 - x - 6}
\end{aligned}$$

EXERCISE
11·3

Divide the rational expressions and give the answers in simplest form.

1. $\dfrac{81 - x^2}{4x + 12} \div \dfrac{9 - x}{x + 3}$

6. $\dfrac{x^2 + 18x + 81}{x^2 - 25} \div \dfrac{x + 9}{3x - 15}$

2. $\dfrac{x^2 - 4x}{x^2 + 2x} \div (x - 4)$

7. $\dfrac{x - 3}{x + 2} \div \dfrac{x^2 + 4x - 21}{3x + 6}$

3. $\dfrac{9x^2}{x^2 - 16} \div \dfrac{3x}{x + 4}$

8. $\dfrac{x^2 - 7x + 12}{x - 1} \div \dfrac{x^2 - 16}{x^2 - 1}$

4. $\dfrac{x^2 - x - 6}{x^2 - 4} \div \dfrac{x^2 - x - 2}{x - 2}$

9. $\dfrac{x^2 - 3x - 10}{x^2 - 4x + 4} \div \dfrac{x - 5}{x - 2}$

5. $\dfrac{2t - 3}{4} \div \dfrac{2t^2 + t - 6}{8t + 16}$

10. $\dfrac{y + 8}{y - 8} \div \dfrac{y^2 - 16y + 64}{y^2 - 64}$

Adding and subtracting rational expressions

Adding and subtracting rational expressions calls on the same skills as adding and subtracting fractions. If the fractions have different denominators, they must be transformed to have a common denominator. Once the denominators are the same, you add or subtract the numerators, and simplify if possible. When you work with rational expressions, because the numerators and denominators are polynomials, that process becomes a little more complicated, but a step-by-step approach will get the job done.

If the fractions have different denominators, as in $\dfrac{5}{x^2-4}-\dfrac{3}{x^2+x-6}$,

- Factor the denominators: $\dfrac{5}{(x+2)(x-2)}-\dfrac{3}{(x+3)(x-2)}$
- Identify any factors common to both denominators, in this case, $x-2$
- Form the lowest common denominator (LCD) from the product of each factor that is common, used once, and any remaining factors of either denominator, for this problem $(x-2)(x+2)(x+3)$
- Transform each fraction by multiplying the numerator and denominator by the same quantity (don't worry about multiplying out the denominator yet)

$$\frac{5}{(x+2)(x-2)}\cdot\frac{(x+3)}{(x+3)}-\frac{3}{(x+3)(x-2)}\cdot\frac{(x+2)}{(x+2)}$$

$$=\frac{5x+15}{(x-2)(x+2)(x+3)}-\frac{3x+6}{(x-2)(x+2)(x+3)}$$

When the fractions have common denominators, add or subtract the numerators. For subtraction, use parentheses around the second numerator to avoid sign errors.

$$\frac{5x+15-(3x+6)}{(x-2)(x+2)(x+3)}=\frac{5x+15-3x-6}{(x-2)(x+2)(x+3)}=\frac{2x+9}{(x-2)(x+2)(x+3)}$$

Finally, factor the numerator if possible, and simplify if possible. Multiply out the denominator at the very last, if necessary.

EXERCISE 11·4

Add or subtract as indicated, and give answers in simplest form.

1. $\dfrac{x+2}{3}+\dfrac{x-3}{2}$

2. $\dfrac{3x-4}{8}-\dfrac{2x+4}{12}$

3. $\dfrac{4}{x+1}+\dfrac{2}{x-2}$

4. $\dfrac{x}{x-5}+\dfrac{3}{x+2}$

5. $\dfrac{x}{4x-3}-\dfrac{x}{4}$

6. $\dfrac{4}{x^2-4}+\dfrac{5}{x+2}$

7. $\dfrac{t^2}{t^2-25}-\dfrac{3t}{3t-15}$

8. $\dfrac{5x}{x^2-9}-\dfrac{5}{x+3}$

9. $\dfrac{x+2}{3x}+\dfrac{3x-1}{x-2}$

10. $\dfrac{x+1}{x+5}+\dfrac{x+2}{x-5}$

11. $\dfrac{a}{a^2-6a+9}-\dfrac{1}{a^2+4a-21}$

12. $\dfrac{5x}{x^2-5x+6}+\dfrac{2}{x-3}-\dfrac{6}{x-2}$

13. $\dfrac{x+3}{x+2}+\dfrac{x-3}{x-2}+\dfrac{x+6}{x^2-4}$

14. $x+\dfrac{4}{x^2-25}+\dfrac{3x-1}{x-5}$

15. $\dfrac{2}{x^2-x-6}-\dfrac{4}{x^2+5x+6}+\dfrac{3}{x^2-9}$

Complex fractions

A complex fraction is one that contains fractions within its numerator or denominator or both. There may be one fraction or several fractions in the numerator or denominator or both.

There are two methods for simplifying complex fractions. The first is to focus your attention initially on the numerator and simplify it as completely as possible and then turn to the denominator and simplify that. The final step in this method is to realize that a fraction is actually a division problem, and divide the numerator by the denominator.

To simplify the complex fraction $\dfrac{\dfrac{1}{x}-\dfrac{1}{2x}}{\dfrac{2}{y}+\dfrac{3}{y^2}}$ by this method, first do the subtraction in the

numerator: $\dfrac{1}{x}-\dfrac{1}{2x}=\dfrac{2}{2x}-\dfrac{1}{2x}=\dfrac{1}{2x}$. Next, turn your attention to the denominator and do that

addition: $\dfrac{2}{y}+\dfrac{3}{y^2}=\dfrac{2y}{y^2}+\dfrac{3}{y^2}=\dfrac{2y+3}{y^2}$. Finally, divide the simplified numerator by the simplified

denominator.

$$\dfrac{1}{2x}\div\dfrac{2y+3}{y^2}=\dfrac{1}{2x}\cdot\dfrac{y^2}{2y+3}=\dfrac{y^2}{4xy+6x}$$

The second method for simplifying complex fractions is often quicker. Find the LCD for all the fractions contained in the numerator and denominator, and then multiply both the numerator

and the denominator by this LCD. For the complex fraction $\dfrac{\dfrac{1}{x}-\dfrac{1}{2x}}{\dfrac{2}{y}+\dfrac{3}{y^2}}$, the LCD would be $2xy^2$.

Multiply the complex fraction by $\dfrac{2xy^2}{2xy^2}$, distribute, and simplify.

$$\dfrac{\dfrac{1}{x}-\dfrac{1}{2x}}{\dfrac{2}{y}+\dfrac{3}{y^2}}\cdot\dfrac{2xy^2}{2xy^2}=\dfrac{2xy^2\left(\dfrac{1}{x}\right)-2xy^2\left(\dfrac{1}{2x}\right)}{2xy\left(\dfrac{2}{y}\right)+2xy^2\left(\dfrac{3}{y^2}\right)}=\dfrac{2y^2-y^2}{4xy+6x}=\dfrac{y^2}{4xy+6x}$$

Simplify each complex fraction.

1. $\dfrac{\dfrac{x}{y}+\dfrac{y}{x}}{\dfrac{x}{y}-\dfrac{y}{x}}$

2. $\dfrac{\dfrac{1}{x}+\dfrac{1}{y}}{\dfrac{1}{xy}}$

3. $\dfrac{6+\dfrac{3}{2x}}{\dfrac{1}{16x}-x}$

4. $\dfrac{\dfrac{x}{x+y}}{1-\dfrac{x}{x+y}}$

5. $\dfrac{1-\dfrac{x}{x-y}}{1+\dfrac{1}{y}}$

6. $\dfrac{\dfrac{1}{x}-1}{\dfrac{1}{x}+1}$

7. $\dfrac{x-2-\dfrac{15}{x}}{x+6+\dfrac{9}{x}}$

8. $\dfrac{x-4+\dfrac{3}{x+4}}{x+4+\dfrac{3}{x-4}}$

9. $\dfrac{\dfrac{16}{y}-y}{\dfrac{8}{y^2}+\dfrac{10}{y}-3}$

10. $\dfrac{6+\dfrac{5x+2}{x^2-1}}{2+\dfrac{1}{x-1}}$

Solving rational equations

There are two methods for solving rational equations. One uses a property of proportions and the other depends on multiplying through to clear fractions.

Cross multiplying

In the cross-multiplication method, you simplify the equation until it is two equal fractions, and then you cross multiply. Cross multiplying gives you an equation that you can solve to find the value of the variable. This method works best when the equation is relatively simple. If you use cross multiplying on rational equations with higher-degree polynomials or just a lot of polynomials in the numerators and denominators, you can end up with an equation that's really difficult to solve, and probably extraneous solutions.

- First, concentrate on the left side of the equation. Perform all indicated operations until the left side is a single fraction.

$$\frac{-20}{x+4}+3=\frac{5}{x+1}-2$$

$$\frac{-20}{x+4}+\frac{3x+12}{x+4}=\frac{5}{x+1}-2$$

$$\frac{3x-8}{x+4}=\frac{5}{x+1}-2$$

- Next concentrate on the right side of the equation. Perform all indicated operations until the right side is a single fraction.

$$\frac{3x-8}{x+4} = \frac{5}{x+1} - 2$$

$$\frac{3x-8}{x+4} = \frac{5}{x+1} - \frac{2x+2}{x+1}$$

$$\frac{3x-8}{x+4} = \frac{-2x+3}{x+1}$$

- Cross multiply and solve the resulting equation.

$$\frac{3x-8}{x+4} = \frac{-2x+3}{x+1}$$

$$(3x-8)(x+1) = (x+4)(-2x+3)$$

$$3x^2 - 5x - 8 = -2x^2 - 5x + 12$$

$$5x^2 - 20 = 0$$

$$5x^2 = 20$$

$$x^2 = 4$$

$$x = \pm 2$$

Always check your solutions in the original equation. Extraneous solutions are not unusual, especially since any values that would make one of the denominators equal to 0 are not in the domain of the equation and therefore can't be solutions.

EXERCISE 11·6

Solve each of the rational equations by cross multiplying.

1. $\dfrac{8}{3x} = 2$

2. $\dfrac{4x-6}{x} = 2$

3. $\dfrac{x}{2} - \dfrac{x}{3} = 4$

4. $\dfrac{a}{a-5} = \dfrac{8}{3}$

5. $\dfrac{1}{x} + \dfrac{1}{3x} = 28$

6. $\dfrac{5}{x} + \dfrac{15}{2x} = \dfrac{5}{4}$

7. $\dfrac{2}{x-1} = \dfrac{3}{x+1}$

8. $\dfrac{1}{x} = \dfrac{1}{x^2-2x}$

9. $\dfrac{7t-3}{4} = \dfrac{5}{2} - \dfrac{t-3}{4}$

10. $\dfrac{3}{2x-8} + \dfrac{1}{4-x} = \dfrac{1}{2}$

Multiplying through by the LCD

The second method for solving rational equations involves clearing, or eliminating, the fractions as quickly as possible. In this method, you multiply every term in the equation by a common denominator to eliminate all the fractions. In the rational equation $\dfrac{1}{x}+3=\dfrac{7}{x}$, multiplying each term by x will give you $1 + 3x = 7$, a simple equation to solve. Of course, many equations will be more complicated, so proceed step by step.

- Factor each of the denominators.

$$\frac{x+1}{x^2+2x}+\frac{3}{x^2+3x+2}=\frac{x+4}{x^2+x}$$

$$\frac{x+1}{x(x+2)}+\frac{3}{(x+1)(x+2)}=\frac{x+4}{x(x+1)}$$

- Determine the LCD of all the algebraic fractions in the equation. The LCD of $x(x + 2)$, $(x + 1)$ $(x + 2)$, and $x(x+1)$ is $x(x + 2)$ $(x + 1)$.
- Multiply both sides of the equation by the LCD, distributing as necessary.

$$x(x+2)(x+1)\left[\frac{x+1}{x(x+2)}\right]+x(x+2)(x+1)\left[\frac{3}{(x+1)(x+2)}\right]=x(x+2)(x+1)\left[\frac{x+4}{x(x+1)}\right]$$

- Cancel as you multiply, and all denominators should disappear.

$$\cancel{x}\,\cancel{(x+2)}(x+1)\left[\frac{x+1}{\cancel{x}\,\cancel{(x+2)}}\right]+x\cancel{(x+2)}\,\cancel{(x+1)}\left[\frac{3}{\cancel{(x+1)}\,\cancel{(x+2)}}\right]=\cancel{x}\,(x+2)\cancel{(x+1)}\left[\frac{x+4}{\cancel{x}\,\cancel{(x+1)}}\right]$$

$$(x+1)(x+1)+x\cdot 3=(x+2)(x+4)$$

- Solve the resulting equation, and check your solution.

$$(x+1)(x+1)+x\cdot 3=(x+2)(x+4)$$
$$x^2+2x+1+3x=x^2+6x+8$$
$$5x+1=6x+8$$
$$-7=x$$

Check:

$$\frac{x+1}{x^2+2x}+\frac{3}{x^2+3x+2}=\frac{x+4}{x^2+x}$$

$$\frac{-7+1}{49+-14}+\frac{3}{49+-21+2}=\frac{-7+4}{49+-7}$$

$$\frac{-6}{35}+\frac{3}{30}=\frac{-3}{42}$$

$$\frac{-36}{210}+\frac{21}{210}=\frac{-15}{210}$$

Solve each rational equation by multiplying through by the LCD.

1. $\dfrac{2}{x} + \dfrac{3}{x-1} = \dfrac{3}{2}$

2. $\dfrac{x+2}{6} - \dfrac{2x+2}{2x+1} = \dfrac{2x+7}{12}$

3. $\dfrac{7}{x-3} + \dfrac{5}{3-x} = \dfrac{2}{3}$

4. $\dfrac{x-3}{x+1} + \dfrac{1}{x-1} = 1$

5. $\dfrac{6x}{x-3} - \dfrac{15}{2x-6} = \dfrac{3}{2}$

6. $\dfrac{5}{y+4} - \dfrac{y}{3y+12} = \dfrac{-20}{3}$

7. $\dfrac{3}{x-5} + \dfrac{17}{20-4x} = \dfrac{5}{8}$

8. $\dfrac{1}{x-5} - \dfrac{5}{3x+15} = \dfrac{8}{x^2-25}$

9. $\dfrac{5}{x+5} - \dfrac{3}{x+8} = \dfrac{7}{x^2+13x+40}$

10. $\dfrac{5}{x+6} - \dfrac{7x-3}{3x^2+17x-6} = \dfrac{3}{3x-1}$

Graphing rational functions

The graphs of rational functions are often discontinuous—that is, the graph is in two or more pieces—because the function is undefined for any value that makes the denominator equal to 0. Simple rational functions have a characteristic two-wing shape called a hyperbola as shown in Figure 11.1, but more complicated rational functions have various graphs. You'll want to make a table of values; there are a few tips that can help you choose useful *x*-values.

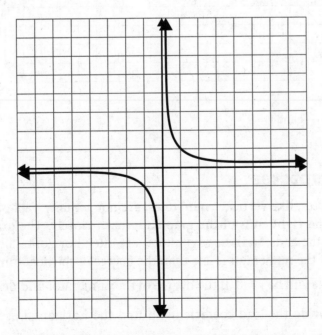

Figure 11.1 The graph of a simple rational function.

Vertical asymptotes

An asymptote is a line that is not part of the graph, but one that the graph approaches closely. When the graph gets close to the vertical asymptote, it curves either upward or downward very steeply so that it looks almost vertical itself. Remember that the graph can get very close to the asymptote but can't touch it.

Vertical asymptotes occur at many of the values of x for which the function is undefined, so before you begin to build a table of values, find the value(s) of x that would make the denominator equal to 0. These will be discontinuities, or breaks, in the graph, and you can expect that as you get near these x-values, the y-values will become very large (positive) or very small (negative).

Choose x-values on both sides of the vertical asymptotes. The function $y = \dfrac{3}{x+1}$ has a vertical asymptote of $x = -1$ and $y = \dfrac{2x+3}{x-2}$ has a vertical asymptote of $x = 2$ as shown in Figure 11.2.

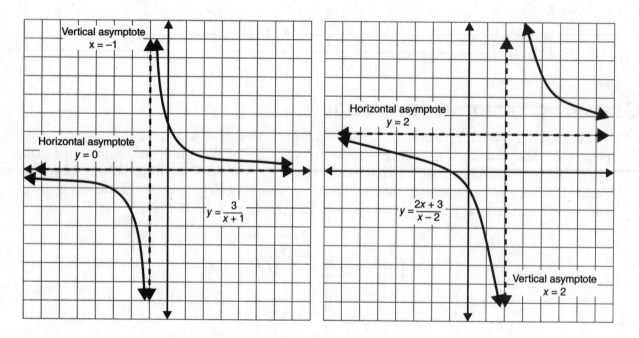

Figure 11.2 Graphs of $y = \dfrac{3}{x+1}$ and $y = \dfrac{2x+3}{x-2}$.

Horizontal asymptotes

Horizontal asymptotes, like vertical asymptotes, are not actually part of the graph, but they are lines that the left and right ends of the graph approach closely. The graphs of simple rationals often flatten out on the ends, so these asymptotes are often horizontal lines.

If the degree of the numerator is less than the degree of the denominator, as in $y = \dfrac{3}{x+1}$, the horizontal asymptote will be $y = 0$. If the degree of the numerator and denominator are the same, as in $y = \dfrac{2x+3}{x-2}$, divide the lead coefficients to find the horizontal asymptote. The horizontal asymptote for $y = \dfrac{2x+3}{x-2}$ is $y = 2$.

Drawing the vertical and horizontal asymptotes as dotted lines before you begin graphing will help you locate the graph.

Graph each rational function. Sketch the vertical and horizontal asymptotes first.

1. $y = \dfrac{1}{x}$

2. $y = \dfrac{1}{x-2}$

3. $y = \dfrac{1}{x+3}$

4. $y = \dfrac{2}{x}$

5. $y = \dfrac{-1}{x}$

6. $y = \dfrac{-3}{x+4}$

7. $y = \dfrac{3x-5}{x-2}$

8. $y = \dfrac{-(x+3)}{x+5}$

9. $y = \dfrac{x}{2x-1}$

10. $y = \dfrac{8x-1}{4x+6}$

Calculator notes #8

The quadratic formula is a critical tool for solving quadratic equations, and should be memorized, but evaluating it often demands the use of a calculator. Sometimes you'll want to leave the expression in simplest radical form, and calculators are not always able to do that, but when the decimal approximation is acceptable, there are a few ways your calculator can help.

With a formula like $x = \dfrac{-b \pm \sqrt{b^2 - 4ac}}{2a}$, typing in the calculation over and over with the different values for each equation can be tedious and prone to errors. Instead try this:

- Type the value of a, press [STO] [ALPHA] A. Repeat to put the value of b into [STO] [ALPHA] B and the value of c into [STO] [ALPHA] C.

```
2→A
                    2
-7→B
                   -7
3→C
                    3
■
```

- Using [ALPHA] A, [ALPHA] B, and [ALPHA] C, type $\left(-B + \sqrt{B^2 - 4AC}\right) / (2A)$ and press [ENTER]. One solution will be displayed.

```
(-B+√(B²-4AC))/(2A⁞
                    3
```

- Cursor up and press ENTER, or press 2nd ENTER, to bring back $\left(-B + \sqrt{B^2 - 4AC}\right)/(2A)$.

$$\left(-B + \sqrt{B^2 - 4AC}\right)/(2A$$
$$3$$
$$3 + \sqrt{B^2 - 4AC}\right)/(2A)$$

- Cursor left, and change the + to − to get $\left(-B - \sqrt{B^2 - 4AC}\right)/(2A)$ and press ENTER. The second solution will be displayed.

$$\left(-B + \sqrt{B^2 - 4AC}\right)/(2A$$
$$3$$
$$\left(-B - \sqrt{B^2 - 4AC}\right)/(2A$$
$$.5$$

- For other problems, store new values for a, b, and c, then recall $\left(-B + \sqrt{B^2 - 4AC}\right)/(2A)$ using 2nd ENTER to get one solution, recall again, and change + to − to get second solution. Repeat as needed.

If you're comfortable doing a little programming, you can create a quadratic formula program on your graphing calculator. It would need to contain a section to ask for the values of a, b, and c, a section to calculate each solution, and a section to display the results.

To create a new program, press PRGM, cursor to NEW, and press ENTER. You'll be asked for a name for the program, and you can use up to eight characters. If you look at the blinking cursor, you'll see it's already in ALPHA mode, and you can type the name using the keys with the appropriate letter above. When you press ENTER, the program will be created, and you'll see a basically blank page. Each new line will start with a colon generated automatically.

The common commands are in menus that you can access by pressing PRGM again. The first menu, CTRL, contains commands for creating loops and directing the flow of the program. The middle menu is commands about input and output, receiving information to use in the program and displaying values, messages, graphs, and tables. The final menu, EXEC, allows your program to call other programs stored on your calculator.

When your program is complete, press 2nd MODE to quit the editing process. The program will be saved automatically. If you need to stop working before you're done, just quit. When you come back, press PRGM and cursor to EDIT. When your program is ready to test or use, press PRGM, choose its name from the list and press ENTER.

The best way to get a quadratic formula program for your calculator is to write it yourself. If that doesn't work for you, there are many versions available on the Internet and even some in textbooks. If you decide to copy one of those, take the time as you enter it to try to understand the logic. Afterward, you may want to try creating one of your own.

Exponential growth and decay

12

An exponent is shorthand for expressing repeated multiplication, a convenient way to condense certain mathematical statements. Large expressions can often be simplified quickly, when written in exponential form, by using the basic rules for exponents. Exponents not only help in arithmetic, but also play an important role in algebra.

When an unknown number is multiplied repeatedly, you get a power of a variable, x^5, and those powers, with their coefficients, can combine into polynomials. In those polynomials, the variable is the base of the power, the number being multiplied. In a polynomial, you have a variable as the base and an integer as the exponent.

When a constant is used as the base and a variable is the exponent, as in 5^x, you have a different kind of expression. Putting the variable in the exponent position, on a constant base, creates an exponential expression or defines an exponential function. In such a function, the base of the power doesn't change. What changes is the number of times it is multiplied. Let's compare a simple polynomial function, $f(x) = x^3$, to an exponential function, $g(x) = 3^x$.

x	0	1	2	3	4
$f(x)$	$0^3 = 0$	$1^3 = 1$	$2^3 = 8$	$3^3 = 27$	$4^3 = 64$

As x increases, the value of x^3 grows, but by a different amount at each step. From $x = 1$ to $x = 2$, there's an increase of 7 in the function value, but from $x = 2$ to $x = 3$, there's a jump of 19. Trying to compare successive values by multiplying doesn't help either, because there's no consistent ratio, $\frac{64}{27} \neq \frac{27}{8} \neq \frac{8}{1}$, and of course, you can't divide 1 by 0.

x	0	1	2	3	4
$g(x)$	$3^0 = 1$	$3^1 = 3$	$3^2 = 9$	$3^3 = 27$	$3^4 = 81$

The exponential function, on the other hand, has a constant ratio of 3. Each term is a power of the base, 3, so each value is 3 times the one before it.

The pattern of an exponential function is also different from a linear pattern. The table below shows some values for $y = 3x$.

x	0	1	2	3	4
y	$3 \times 0 = 0$	$3 \times 1 = 3$	$3 \times 2 = 6$	$3 \times 3 = 9$	$3 \times 4 = 12$

In this function, there is no constant ratio, but there is a constant difference. For each increase of 1 unit in x, y increases by 3, the slope of the linear equation. In a linear equation, you see a constant difference between successive terms, equal to the slope. In a power equation, like $y = x^3$, you see neither a constant difference nor a constant ratio, but in an exponential equation, there is a constant ratio of successive terms. You can use those differences to help distinguish among the types of equations.

The three equations we looked at ($y = 3x$, $y = x^3$, and $y = 3^x$) are simple examples of their types, and each type has its own standard form. Exponential functions have a standard form of $y = ab^x$, where a and b are constants and b is greater than zero. If we allowed the base b to be negative, and if x were to take a value like $\frac{1}{2}$ that represents a root, we might find ourselves trying to take the square root of a negative number. There is no such real number, so requiring b>0 eliminates the problem.

The value of a is a multiplier, often representing a beginning value $b^0 = 1$ for any non-zero value of b, but 1 may not always be where you want to start. Multiplying by a lets you change that. However, that extra multiplier can make it harder to recognize an exponential relationship. Let's compare two functions called y_1 and y_2.

x	0	1	2	3	4
y_1	1	2	4	8	16
y_2	3	6	12	24	48

It's easy to recognize the values of y_1 as an exponential function. They're simply powers of two, so the equation is $y_1 = 2^x$. The function y_2 is not as easy to identify. Start by looking at ratios $\frac{6}{3} = \frac{12}{6} = \frac{24}{12} = \frac{48}{24} = 2$. The ratio of successive terms is always 2, which says that this is an exponential function and the base of the power is two. Now take $y = ab^x$, replace b with 2, and then plug in any corresponding x and y from the table. You'll be able to solve for a. The easiest order pair to choose would be (0, 3) but any pair will do so let's use (3, 24); $24 = a(2^3)$ becomes $24 = 8a$, and dividing by 8 gives you $a = 3$. The equation is $y_2 = 3(2^x)$.

EXERCISE
12·1

Each of the tables below shows values of either a linear function, $(y = mx + b)$, a power function, $(y = ax^n)$, or an exponential function $(y = ab^x)$. Identify which type of function is represented in each table.

1.

x	1	2	3	4	5
y	10	100	1,000	10,000	100,000

2.

x	1	2	3	4	5
y	10	14	18	22	26

3.

x	3	4	5	6	7
y	18	32	50	72	98

4.

x	3	4	5	6	7
y	$\frac{5}{8}$	$\frac{5}{16}$	$\frac{5}{32}$	$\frac{5}{64}$	$\frac{5}{128}$

5.

x	6	7	8	9	10
y	11	24	39	56	75

Each of the tables below represents an exponential function $\left(y = ab^x\right)$. Identify the values of a and b.

6.

x	0	1	2	3	4
y	2	6	18	54	162

7.

x	0	1	2	3	4
y	3	6	12	24	48

8.

x	1	2	3	4	5
y	1.5	4.5	13.5	40.5	121.5

9.

x	1	2	3	4	5
y	$-\dfrac{2}{3}$	$-\dfrac{2}{9}$	$-\dfrac{2}{27}$	$-\dfrac{2}{81}$	$-\dfrac{2}{243}$

10.

x	2	3	4	5	6
y	200	2,000	20,000	200,000	2,000,000

Evaluating exponential functions

Exponential functions grow quickly, even when the base is small, and the base is not restricted to positive integers, so fractions or decimals could be used as bases, as long as they're positive. For these reasons, a calculator is handy when you're evaluating exponential expressions and functions.

Calculate carefully when you have a multiplier, and watch the placement of parentheses, especially on a calculator. Remember the order of operations says exponents before multiplication, so raise b—and only b—to the xth power, and then multiply by a. For the exponential function $f(x) = 6\left(\frac{1}{2}\right)^x$, the value of $f(4) = 6\left(\frac{1}{2}\right)^4 = 6\left(\frac{1}{16}\right) = \frac{3}{8}$. If you mistakenly multiply 6 times $\frac{1}{2}$ first, you'll end up with $3^4 = 81$, a very different (and incorrect) result.

Given the y-value for an exponential function, it's more challenging to find the value of x. Many exponential equations require knowledge beyond the scope of an Algebra I course, but some can be solved with skills you have, if the numbers are friendly.

If you're asked to solve $5^x = 25$, you likely just recognized that $5^2 = 25$, so x must be 2. To solve $-7\left(3^x\right) = -189$, first divide both sides by –7, to get $3^x = 27$. Then previous experience (or a little trial and error) will tell you that $3^3 = 27$, so $x = 3$. But you can also see that $2^x = 17$ doesn't yield to this method. It will surrender when you study logarithms.

EXERCISE 12·2

For questions 1 – 6 find the value of y for the given value of x.

1. $y = 2\left(3^x\right)$ and $x = 4$.

2. $y = \frac{1}{2}\left(4^x\right)$ and $x = 2$.

3. $y = -1\left(5^x\right)$ and $x = 3$.

4. $y = 0.25\left(6^x\right)$ and $x = 2$.

5. $y = 3\left(10^x\right)$ and $x = 6$.

6. $y = -2\left(0.1^x\right)$ and $x = 2$.

For questions 7 – 10, solve each equation.

7. $2\left(5^x\right) = 50$

8. $-4\left(3^x\right) = -36$

9. $7\left(2^x\right) = 56$

10. $\frac{1}{2}\left(10^x\right) = 5{,}000$

Compound interest

When you invest a sum of money, you expect that you'll get your money back plus an additional amount, called interest. If you borrow money, you expect to repay the loan plus interest. The amount originally invested or borrowed is the principal, and the amount of interest earned or paid is determined by finding a percentage of the principal.

The interest paid on a loan or an investment can be calculated either as simple or compound interest. Simple interest is a one-time calculation: the amount loaned or invested times the interest rate per year times the number of years, or $I = Prt$. Compound interest is calculated at regular intervals—annually, semiannually, quarterly, monthly, or daily—and the interest for that period is added to the principal, and then earns interest along with the principal. If you invest $1000 at 5% per year for 2 years, simple interest would be $I = 1000 \times 0.05 \times 2 = \$100$, for a total of $1100. If you invest the same $1000 at 5% per year for 2 years compounded annually, interest would be calculated at the end of the first year: $I = 1000 \times 0.05 \times 1 = \$50$, and that $50 would be added to the original $1000. The interest for the second year would be calculated on $1050: $I = 1050 \times 0.05 \times 1 = \$52.50$. After 2 years you will have earned $102.50 for a total of $1102.50.

After the first year, you had your original $1000 plus 5%, or 105%, of your original investment. After 2 years, you had 105% of 105% of your original investment, or $\$1000(1.05)^2$. When interest is compounded annually, the total value of the investment is $A = P(1+r)^t$, where P is the principal or original investment, r is the rate of interest per year (converted to a decimal), and t is the number of years.

If the interest is compounded more than once a year, you don't get the whole year's worth of interest at every interval. Instead, the annual rate is divided by the number of times per year the calculation is done. While this might look like it would reduce the interest, remember that interest will be compounded more frequently and each time it is compounded, the amount of money earning interest grows.

For compound interest, use the formula $A = P\left(1+\dfrac{r}{n}\right)^{nt}$, where P is the principal or original investment, r is the rate of interest per year (converted to a decimal), n is the number of times per year interest is compounded, and t is the number of years.

EXERCISE
12·3

Calculate the value of each investment after the specified time, when invested as described.

1. An investment of $5000, at 3% per year, compounded annually, for 2 years

2. An investment of $10,000, at 8% per year, compounded semiannually, for 5 years

3. An investment of $2000, at 4% per year, compounded quarterly, for 10 years

4. An investment of $2500, at 5% per year, compounded monthly, for 4 years

5. An investment of $100,000, at 10% per year, compounded quarterly, for 12 years

6. An investment of $4000, at 5.5% per year, compounded monthly, for 8 years

7. An investment of $7500, at 2.5% per year, compounded semiannually, for 5 years

8. An investment of $3000, at 9% per year, compounded annually, for 8 years

9. An investment of $15,000, at 12% per year, compounded quarterly, for 6 years

10. An investment of $25,000, at 6% per year, compounded annually, for 15 years

Exponential growth and decay

Compound interest is a common example of exponential growth. The principal amount grows over time because it's multiplied by powers of $1 + r$ or $1 + \dfrac{r}{n}$. When the base of a power is greater than 1, the power will grow as the exponent increases, so equations of the form $y = ab^x$ represent exponential growth when b is greater than 1.

In contrast, when b is less than 1 but greater than 0, the equation represents exponential decay, a decreasing quantity. If the amount of math George knows decreases 2% per week over summer vacation, after 1 week he'll know 98% (or $1 - 0.02$) of what he knew the week before. After 3 weeks, he'll know 98% of 98% of 98%, or $(1 - 0.02)^3$. When the rate of increase or decrease is given, it may be helpful to rewrite the equation as $y = a(1+r)^x$ for increase or $y = a(1-r)^x$ for decrease.

EXERCISE
12·4

Tell whether each equation represents exponential growth or exponential decay.

1. $y = 400(1 + 0.03)^x$

2. $y = 30,000(1 - 0.01)^x$

3. $y = 42\left(\dfrac{3}{5}\right)^x$

4. $y = 950\left(\dfrac{7}{3}\right)^x$

5. $y = 2.05(0.6)^x$

Identify each situation as growth or decay, and evaluate the result.

6. A colony of bacteria is created with 200 bacteria, and the population doubles every hour. Find the population 1 day (24 h) later.

7. A patient is given an injection of 250 mg of a drug. Each hour, as the body metabolizes the drug, the level in the bloodstream is reduced by 20%. What is the level in the bloodstream 4 h later?

8. A city had a population of 250,000 in 2008, and the population was increasing by 11% per year. What would the population be in 2012?

9. A new car is purchased for $24,000. The car depreciates (loses value) at 12% per year. How much is the car worth 3 years later?

10. A county had 45,000 acres of forested land in 1996, but that acreage was decreasing at 5% per year. How many acres of forested land remained in the county in 2000?

Graphing the exponential functions

Graphs of exponential functions have a characteristic shape, almost flat on one end and very steep on the other. The flat end approaches a horizontal asymptote, a horizontal line that the graph comes very close to but doesn't touch. The graph of an exponential growth rises sharply on the right and approaches a horizontal asymptote on the left. Exponential decay graphs are reflected, reversed left to right, so they're falling steeply from left to right and flattening out on the right end (see Figure 12.1).

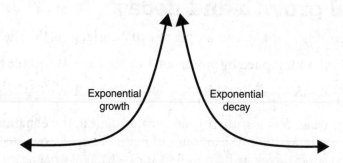

Figure 12.1 Exponential growth and decay.

Plotting a few key points can help you shape the graph of an exponential equation. Always look for the y-intercept. If $y = ab^x$, when $x = 0$, $b^0 = 1$, so the y-intercept will be a. Plugging in 1 and -1 for x will give you two more points that will easily set the shape: $(1, ab)$ and $\left(-1, \dfrac{a}{b}\right)$. The graph of $y = 3(2)^x$ has a y-intercept of $(0, 3)$ and passes through the points $(1, 6)$ and $\left(-1, \dfrac{3}{2}\right)$ as shown in Figure 12.2.

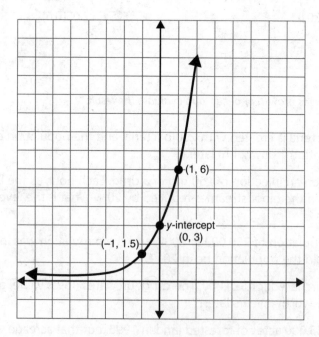

Figure 12.2 Graph of exponential growth.

Graph each exponential function by making a table of values and plotting points.

1. $y = 2(1.5)^x$

2. $y = 4(.75)^x$

3. $y = 2^x$

4. $y = \left(\dfrac{1}{3}\right)^x$

5. $y = 3(2)^x$

6. $y = -1(3)^x$

7. $y = 3\left(\dfrac{1}{2}\right)^x - 1$

8. $y = -2(1.5)^x + 4$

9. $y = 5(.9)^x + 2$

10. $y = 10(1.1)^x - 3$

Calculator notes #9

In first-year algebra, your tools for solving exponential equations are limited. You can solve an exponential equation if it's possible to express both sides of the equation as powers of the same base, but if that's not the case, the best you can do is to take a guess. If you tried to solve the equation $5\left(2^x\right) + 3 = 163$, you could subtract 3 and divide by 5 and get to $2^x = 32$. Because 32 is a power of two, you can express that as $2^x = 2^5$ and conclude that $x = 5$. On the other hand, if the original equation were $5\left(2^x\right) + 3 = 165$, the solution would look like this:

$$5\left(2^x\right) + 3 = 165$$

$$5\left(2^x\right) = 162$$

$$2^x = 32.4$$

No integer power of two is equal to 32.4, so you can only say x is a little more than 5.

With your graphing calculator, however, you can use the intersect feature to get a better estimate.

- Graph $Y1 = 2^x$ and $Y2 = 32.4$. Adjust the window so that you can see where the graphs intersect.

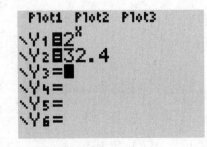 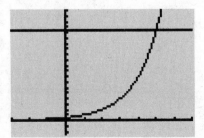

- Press ⌐2nd⌐ ⌐TRACE⌐. Choose 5: intersect.
- Press ⌐ENTER⌐ to mark the first curve.
- Press ⌐ENTER⌐ to mark the second curve.

- Move your cursor near the point of intersection if you wish, and press ENTER.

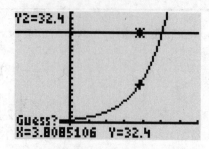

- The solution is displayed at the bottom of the screen.

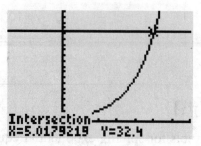

Matrix algebra

The word *matrix* is used in many contexts with different meanings. In mathematics, a matrix is a rectangular arrangement of numbers. Numbers are often organized into matrices because they represent similar pieces of information or because the same calculation must be performed on all of them. On the coordinate plane shown in Figure 13.1, the triangle has vertices at the points (1, 2), (−1, −1), and (2, −1). The *x*-coordinates are on the top row, and the *y*-coordinates on the bottom. Each column represents a point. It can be represented by the matrix

$$\begin{bmatrix} 1 & -1 & 2 \\ 2 & -1 & -1 \end{bmatrix}$$

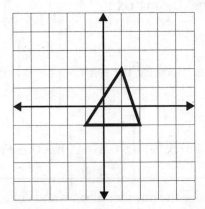

Figure 13.1 The vertices of a triangle can be organized in a matrix.

Matrices are generally enclosed in square brackets, although sometimes other enclosures are used. Any rectangular arrangement may be considered a matrix, even if it is not enclosed (see Figure 13.2).

$$\begin{bmatrix} 2 & 8 & 3 & 6 & 1 \\ 8 & 4 & 2 & 6 & 5 \end{bmatrix}$$

$$\begin{pmatrix} 1 & 0 \\ 5 & -1 \end{pmatrix}$$

$$\begin{array}{c|c} 3 & 7 \\ \hline 6 & 9 \\ \hline 2 & 5 \end{array}$$

$$\begin{array}{ccc} 4 & 7 & 2 \\ 9 & 8 & 4 \\ 1 & 0 & 9 \end{array}$$

Figure 13.2 Matrices can be written in different formats.

Rows and columns

The matrix as a whole can be named by a single capital letter, and the individual numbers within the matrix are called elements. We can say $A = \begin{bmatrix} 4 & 7 & 2 \\ 9 & 8 & 4 \\ 1 & 0 & 9 \end{bmatrix}$ or $M = \begin{bmatrix} 1 & 0 \\ 5 & -1 \end{bmatrix}$. Every matrix is organized into rows, which are horizontal lines of elements, and columns, which are vertical stacks of numbers. The matrix $\begin{bmatrix} 2 & 8 & 3 & 6 & 1 \\ 8 & 4 & 2 & 6 & 5 \end{bmatrix}$ has two rows, each containing five elements. The matrix $\begin{bmatrix} 3 & 7 \\ 6 & 9 \\ 2 & 5 \end{bmatrix}$ has three rows, each containing two elements. The matrix $\begin{bmatrix} 2 & 8 & 3 & 6 & 1 \\ 8 & 4 & 2 & 6 & 5 \end{bmatrix}$ has five columns of two elements, and the matrix $\begin{bmatrix} 3 & 7 \\ 6 & 9 \\ 2 & 2 \end{bmatrix}$ has two columns of three elements.

The dimension, or order, of a matrix is a description of its size, giving first the number of rows and then the number of columns. The matrix $\begin{bmatrix} 2 & 8 & 3 & 6 & 1 \\ 8 & 4 & 2 & 6 & 5 \end{bmatrix}$ is a 2×5 matrix, meaning that it has two rows and five columns. Because the matrix $\begin{bmatrix} 3 & 7 \\ 6 & 9 \\ 2 & 5 \end{bmatrix}$ has three rows and two columns, we say its dimension is 3×2. Matrix $A = \begin{bmatrix} 4 & 7 & 2 \\ 9 & 8 & 4 \\ 1 & 0 & 9 \end{bmatrix}$ has three rows and three columns, so it has dimension, or order, 3×3. Because the number of rows and columns are the same in matrix A, we say that matrix A is square. Matrix $M = \begin{bmatrix} 1 & 0 \\ 5 & -1 \end{bmatrix}$ is also a square matrix and its dimension is 2×2.

A matrix with only one row is called a row matrix and a matrix with only one column is called a column matrix. Row matrices and column matrices are sometimes referred to as vectors.

EXERCISE
13·1

Give the dimension of each matrix.

1. $\begin{bmatrix} 9 & 7 & 3 & 9 \\ 5 & 3 & 2 & 0 \end{bmatrix}$

2. $\begin{bmatrix} 1 & 0 & 2 & 9 \end{bmatrix}$

3. $\begin{bmatrix} -1 & 9 \\ -6 & -4 \\ 7 & 0 \end{bmatrix}$

5. $\begin{bmatrix} 1 & 0 & 0 \\ 0 & 1 & 0 \\ 0 & 0 & 1 \end{bmatrix}$

4. $\begin{bmatrix} 4 \\ 6 \end{bmatrix}$

6. A small sporting goods shop keeps records of the types of purchases made by its customers. These records are organized into categories of equipment, clothing, accessories, and books, and then each category is divided by sport. Organize the records from a typical day, below, into a matrix.

    Equipment: 24 tennis, 15 golf, 2 volleyball, 7 softball, 3 basketball

    Clothing: 5 tennis, 2 golf, 1 basketball, 2 softball

    Accessories: 3 golf, 1 volleyball, 5 softball

    Books: 2 tennis, 12 golf, 1 basketball

Addition and subtraction

The sporting goods store mentioned in the last section would certainly want to combine the information gathered on one day with information from other days to see total sales for a week, a month, a quarter, or a year. This could be accomplished by adding the individual numbers, but entering the information into matrices, as you did in the exercise, can simplify the process. With the aid of calculators or computers, the computation is streamlined, becoming one operation rather than many, but even if the work must be done manually, the matrix structure clarifies the task and helps to prevent errors.

Matrices to be added or subtracted must be of the same dimensions. Only matrices with identical dimensions can be added or subtracted. If the sizes of the matrices are not the same, the addition or subtraction cannot be performed.

Logically, it is also important to consider what the matrices represent. It would make little sense to add a matrix showing sales of sporting goods to a matrix containing calorie counts. What could the total possibly represent? Even if the differences between the matrices are not so dramatic, care must be taken to assure that the calculation is sensible.

If two matrices have the same dimension, they can be added by simply adding the corresponding elements. If the matrices did not have the same dimension, some elements would not have partners, and it would be impossible to complete the addition properly.

To add the matrix $\begin{bmatrix} 8 & 6 & 2 & 9 \\ 7 & 1 & 5 & 0 \end{bmatrix}$ to the matrix $\begin{bmatrix} 9 & 3 & 7 & 2 \\ 0 & 4 & 1 & 8 \end{bmatrix}$, we create a new matrix of the same dimension, and fill it with the sums of the corresponding elements.

$$\begin{bmatrix} 8 & 6 & 2 & 9 \\ 7 & 1 & 5 & 0 \end{bmatrix} + \begin{bmatrix} 9 & 3 & 7 & 2 \\ 0 & 4 & 1 & 8 \end{bmatrix} = \begin{bmatrix} 8+9 & 6+3 & 2+7 & 9+2 \\ 7+0 & 1+4 & 5+1 & 0+8 \end{bmatrix} = \begin{bmatrix} 17 & 9 & 9 & 11 \\ 7 & 5 & 6 & 8 \end{bmatrix}$$

The process of subtracting matrices is similar to that of adding matrices. Matrices must be of the same dimension, and corresponding elements are subtracted. To subtract the matrix $\begin{bmatrix} 8 & 6 & 2 & 9 \\ 7 & 1 & 5 & 0 \end{bmatrix}$ from the matrix $\begin{bmatrix} 9 & 3 & 7 & 2 \\ 0 & 4 & 1 & 8 \end{bmatrix}$, form a new matrix of the same dimension and fill it with the differences of the corresponding elements.

$$\begin{bmatrix} 8 & 6 & 2 & 9 \\ 7 & 1 & 5 & 0 \end{bmatrix} - \begin{bmatrix} 9 & 3 & 7 & 2 \\ 0 & 4 & 1 & 8 \end{bmatrix} = \begin{bmatrix} 8-9 & 6-3 & 2-7 & 9-2 \\ 7-0 & 1-4 & 5-1 & 0-8 \end{bmatrix} = \begin{bmatrix} -1 & 3 & -5 & 7 \\ 7 & -3 & 4 & -8 \end{bmatrix}$$

Just as in standard arithmetic, order is significant in subtraction. You know that $7 - 3 \neq 3 - 7$. The first equals 4, while the second gives -4. Similarly, the result of $\begin{bmatrix} 8 & 6 & 2 & 9 \\ 7 & 1 & 5 & 0 \end{bmatrix} - \begin{bmatrix} 9 & 3 & 7 & 2 \\ 0 & 4 & 1 & 8 \end{bmatrix}$ is not the same as $\begin{bmatrix} 9 & 3 & 7 & 2 \\ 0 & 4 & 1 & 8 \end{bmatrix} - \begin{bmatrix} 8 & 6 & 2 & 9 \\ 7 & 1 & 5 & 0 \end{bmatrix}$.

Changing the order of subtraction changed the sign of each element in the final matrix.

When you learned to subtract integers, you probably were taught to "add the opposite" or to "change the sign and add." These rules told you, for example, that $7 - (-3) = 7 + (+3)$. When subtracting matrices, you can apply a similar rule.

If $A = \begin{bmatrix} -1 & 3 \\ 5 & 0 \\ 2 & -5 \end{bmatrix}$ and $B = \begin{bmatrix} 4 & -3 \\ -2 & 1 \\ -4 & 5 \end{bmatrix}$, then the problem can be expressed as $\begin{bmatrix} -1 & 3 \\ 5 & 0 \\ 2 & -5 \end{bmatrix} + \begin{bmatrix} -4 & 3 \\ 2 & -1 \\ 4 & -5 \end{bmatrix}$. Each element of the second matrix has been changed to its opposite, and you add instead of subtracting. Only the second matrix, the one following the minus sign, is changed.

When matrix addition or subtraction is used in applications, it is important to be certain that the matrices are organized in ways that assure that the operation is sensible. Attempting to add the matrix

	Tennis	Golf	Volleyball	Softball	Basketball
Equipment	23	15	2	7	3
Clothing	5	2	0	2	1
Accessories	0	3	1	5	0
Books	2	12	0	0	1

to the matrix

	Tennis	Golf	Volleyball	Softball	Basketball
Accessories	16	7	8	11	3
Books	5	0	7	3	8
Clothing	18	0	28	8	3
Equipment	4	6	0	0	1

would produce numbers with little meaning, since the categories appear in different orders in each matrix. One of the matrices should first be reorganized so that corresponding elements represent like quantities.

Add, if possible.

1. $\begin{bmatrix} 5 & 3 \\ 2 & 7 \end{bmatrix} + \begin{bmatrix} 1 & 8 \\ 5 & 3 \end{bmatrix}$

2. $\begin{bmatrix} 2 & -1 & 9 \\ 4 & -2 & 6 \end{bmatrix} + \begin{bmatrix} -1 & 4 & 6 \\ 5 & 3 & -9 \end{bmatrix}$

3. $[9 \ \ 8 \ \ 3 \ \ 6 \ \ 2 \ \ 6 \ \ 0] + [1 \ \ -2 \ \ 5 \ \ -3 \ \ 0 \ \ 4 \ \ 3]$

4. $\begin{bmatrix} 2 & 4 & 5 & 2 & 7 \\ 4 & 8 & 4 & 3 & 2 \\ 9 & 7 & 5 & 7 & 0 \end{bmatrix} + \begin{bmatrix} 3 & 2 & 5 & 7 \\ 4 & 8 & 0 & 4 \\ 3 & -5 & 8 & -4 \end{bmatrix}$

5. $\begin{bmatrix} \dfrac{1}{5} \\ \dfrac{1}{2} \end{bmatrix} + \begin{bmatrix} \dfrac{3}{5} \\ \dfrac{1}{3} \end{bmatrix}$

Subtract, if possible.

6. $\begin{bmatrix} 7.24 & 6.83 & 4.81 & 9.33 \\ 5.39 & 7.48 & 3.82 & 7.29 \end{bmatrix} - \begin{bmatrix} 6.23 & 2.71 & 1.77 & 7.21 \\ 1.03 & 3.28 & 0.28 & 7.01 \end{bmatrix}$

7. $\begin{bmatrix} 16 & 8 & 9 \\ 7 & 14 & 6 \\ 3 & 23 & 15 \end{bmatrix} - \begin{bmatrix} 8 & 3 & 5 \\ 2 & 11 & 3 \\ 0 & 15 & 7 \end{bmatrix}$

8. $[3 \ \ 8 \ \ 6] - \begin{bmatrix} 2 \\ 0 \\ 1 \end{bmatrix}$

9. $\begin{bmatrix} 2 & -8 \\ -9 & 6 \\ 0 & 3 \end{bmatrix} - \begin{bmatrix} 8 & -3 \\ 7 & 4 \\ -1 & 9 \end{bmatrix}$

10. $\begin{bmatrix} \dfrac{3}{4} \\ \dfrac{5}{6} \end{bmatrix} - \begin{bmatrix} \dfrac{1}{4} \\ \dfrac{1}{6} \end{bmatrix}$

Scalar multiplication

In matrix arithmetic, there are two types of multiplication, scalar multiplication and matrix multiplication. As the name suggests, the latter involves the multiplication of a matrix by a matrix, and we will consider that in the next section. Scalar multiplication, on the other hand, is the multiplication of a single number times a matrix. The single number is called a scalar.

In the scalar multiplication $3\begin{bmatrix} 1 & 7 & -2 \\ 0 & 5 & -1 \end{bmatrix}$, multiplying the 2×3 matrix by the scalar 3 has the effect of adding three copies of the matrix.

$$3\begin{bmatrix} 1 & 7 & -2 \\ 0 & 5 & -1 \end{bmatrix} = \begin{bmatrix} 1 & 7 & -2 \\ 0 & 5 & -1 \end{bmatrix} + \begin{bmatrix} 1 & 7 & -2 \\ 0 & 5 & -1 \end{bmatrix} + \begin{bmatrix} 1 & 7 & -2 \\ 0 & 5 & -1 \end{bmatrix}$$

Because scalar multiplication represents repeated addition, we can easily anticipate the end result of the process.

$$3\begin{bmatrix} 1 & 7 & -2 \\ 0 & 5 & -1 \end{bmatrix} = \begin{bmatrix} 1 & 7 & -2 \\ 0 & 5 & -1 \end{bmatrix} + \begin{bmatrix} 1 & 7 & -2 \\ 0 & 5 & -1 \end{bmatrix} + \begin{bmatrix} 1 & 7 & -2 \\ 0 & 5 & -1 \end{bmatrix} = \begin{bmatrix} 3 & 21 & -6 \\ 0 & 15 & -3 \end{bmatrix}$$

Focusing on the beginning and end of this process allows us to find the common shortcut for scalar multiplication.

$$3\begin{bmatrix} 1 & 7 & -2 \\ 0 & 5 & -1 \end{bmatrix} = \begin{bmatrix} 3 & 21 & -6 \\ 0 & 15 & -3 \end{bmatrix} = \begin{bmatrix} 3 \cdot 1 & 3 \cdot 7 & 3 \cdot (-2) \\ 3 \cdot 0 & 3 \cdot 5 & 3 \cdot (-1) \end{bmatrix}$$

To multiply a matrix by a scalar, multiply each element of the matrix by the scalar.

The rule for scalar multiplication may remind you of the process you learned as the distributive law. While the scalar certainly seems to be distributed over the matrix, there is a significant difference between the two ideas. The distributive property distributes multiplication over addition (or subtraction), assuring us that $c(a+b)=ca+cb$. In scalar multiplication, the elements of the matrix are not added to one another, either before or after the multiplication.

When scalar multiplication is combined with addition and subtraction, the familiar order of operations will apply. First perform any scalar multiplication, then add or subtract from left to right. As always, matrices must have the same dimension if addition or subtraction is to be performed.

$$2\begin{bmatrix} 4 & -1 \\ 3 & 6 \end{bmatrix} + 3\begin{bmatrix} -5 & 2 \\ -2 & 1 \end{bmatrix} = \begin{bmatrix} 8 & -2 \\ 6 & 12 \end{bmatrix} + \begin{bmatrix} -15 & 6 \\ -6 & 6 \end{bmatrix} = \begin{bmatrix} -7 & 4 \\ 0 & 18 \end{bmatrix}$$

Applications to coordinate geometry

The name scalar multiplication comes from the fact that the single number multiplying the matrix represents a scale factor, an indication of a proportional change in the size. This root meaning of the term is simplest to see when we consider a figure in a coordinate plane represented by a matrix of its coordinates as shown in Figure 13.3.

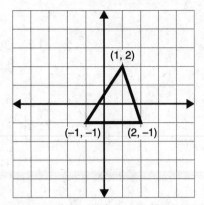

Figure 13.3 The vertices of the triangle can be represented in a matrix.

If the triangle shown on the grid at the left is represented by the matrix $\begin{bmatrix} 1 & -1 & 2 \\ 2 & -1 & -1 \end{bmatrix}$ and we multiply that matrix by a scalar factor of 2, we produce a new matrix.

$$2\begin{bmatrix} 1 & -1 & 2 \\ 2 & -1 & -1 \end{bmatrix} = \begin{bmatrix} 2 & -2 & 4 \\ 4 & -2 & -2 \end{bmatrix}$$

If we graph the points represented by this new matrix, we find that they form the vertices of a triangle similar to the original, but with sides twice as long. Each vertex of the image triangle is twice as far from the origin as the corresponding vertex of the original triangle as shown in Figure 13.4.

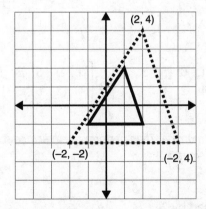

Figure 13.4 The new triangle is twice the size.

Multiply.

1. $5\begin{bmatrix} -3 & 5 \\ 7 & -1 \end{bmatrix}$

2. $-4[1.25 \quad 3.75 \quad -2.5 \quad 4]$

3. $\dfrac{1}{2}\begin{bmatrix} 42 & 36 & 78 \\ 94 & 14 & 17 \end{bmatrix}$

4. $-1.2\begin{bmatrix} 4 & 0.5 \\ -3 & 2 \\ 0 & -5 \end{bmatrix}$

5. $2.50\begin{bmatrix} 16 \\ 18 \\ 21 \\ 8 \end{bmatrix}$

Perform the indicated operations, if possible.

6. $3\begin{bmatrix} 1 & 3 & 4 \\ 5 & 2 & 6 \\ 2 & 1 & 3 \end{bmatrix} + 2\begin{bmatrix} 2 & 6 & 3 \\ 4 & 5 & 1 \\ 0 & 6 & 3 \end{bmatrix}$

7. $-2\begin{bmatrix} 3 & 2 \\ 5 & 1 \\ 0 & 3 \end{bmatrix} + 7\begin{bmatrix} 2 & 5 \\ -1 & -2 \\ 3 & 0 \end{bmatrix}$

8. $5\begin{bmatrix} 1 & 5 \\ 2 & 4 \end{bmatrix} + 2\begin{bmatrix} 1 & 7 & 4 \\ 2 & 3 & 1 \end{bmatrix}$

9. $-3\begin{bmatrix} 1 \\ 2 \\ 0 \end{bmatrix} + 3\begin{bmatrix} -2 \\ 0 \\ -2 \end{bmatrix}$

10. $2\begin{bmatrix} 1 & 0 & 5 & 3 & 5 \\ 4 & 2 & 6 & 3 & 1 \end{bmatrix} - 5\begin{bmatrix} 1 & 0 & 5 & 3 & 5 \\ 4 & 2 & 6 & 3 & 1 \end{bmatrix} + 3\begin{bmatrix} 1 & 0 & 5 & 3 & 5 \\ 4 & 2 & 6 & 3 & 1 \end{bmatrix}$

Graph the points represented by matrix A. Then perform the indicated scalar multiplication and graph the points represented by the answer matrix.

11. $A = \begin{bmatrix} 4 & 1 & -3 \\ 1 & 5 & 0 \end{bmatrix}$ multiplied by 2

12. $A = \begin{bmatrix} 3 & 0 & -1 \\ -1 & 2 & -2 \end{bmatrix}$ multiplied by 3

13. $A = \begin{bmatrix} 4 & 2 & -4 \\ 2 & 8 & 0 \end{bmatrix}$ multiplied by $\dfrac{1}{2}$

14. The four classes in a high school compete in a fundraising event in which they sell T-shirts for \$12.99 each. Seniors sold 172 shirts, juniors 88, sophomores 106, and freshmen 42. Organize the sales numbers into a matrix, and use scalar multiplication to find the amount of money raised by each class.

Matrix multiplication

When a matrix is multiplied by a scalar, every element of the matrix is multiplied by that same value. While this is useful in some situations, many times the necessary calculations are more complicated.

In the last exercise of the previous section, you created a small matrix and multiplied it by the scalar \$12.99. Imagine, however, that the classes sold both T-shirts and sweatshirts. For each class, you would need to record both the number of T-shirts sold and the number of sweatshirts sold. This increases the size of your matrix, but also introduces another problem. It is unlikely that the T-shirts and the sweatshirts would sell for the same price. Sweatshirts would probably cost more, perhaps \$15.99. Now you have a situation in which some elements of the matrix need to be multiplied by \$12.99 while others must be multiplied by \$15.99. The solution is matrix multiplication.

Two matrices can be multiplied only if the number of elements in each row of the first matrix is equal to the number of elements in each column of the second. If $A = \begin{bmatrix} 1 & 3 & 2 \\ -2 & 0 & 4 \end{bmatrix}$ and $B = \begin{bmatrix} 1 & 0 & 3 & -2 \\ 2 & 1 & -4 & 3 \\ 5 & 0 & -3 & 1 \end{bmatrix}$, we can multiply $A \cdot B$ because there are three elements of each row of A and three elements of each column of B. In other words, the number of columns in the first matrix must be equal to the number of rows in the second. Matrix A has two rows and three columns, and matrix B has three rows and four columns.

If we write the dimensions of matrix A and then the dimensions of matrix B, we can see a simple way to tell if the multiplication is possible (see Figure 13.5).

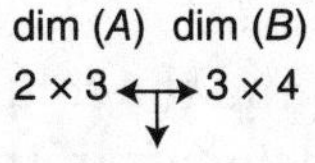

dim (A) dim (B)

$2 \times 3 \longleftrightarrow 3 \times 4$

Figure 13.5 If dimensions match, the multiplication is possible.

If the dimensions do not match, the multiplication cannot be performed (see Figure 13.6). The need for matching dimensions makes the order of multiplication important. While we have seen that it is possible to multiply $A \cdot B$, it is not possible to perform the multiplication $B \cdot A$.

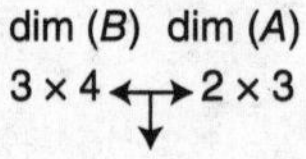

dim (*B*) dim (*A*)

$3 \times 4 \leftrightarrow 2 \times 3$

Figure 13.6 These numbers do not match, so the multiplication is not possible.

Looking at the dimensions can give us another piece of useful information as well. The remaining numbers tell us the dimension of the product matrix that will result. When we multiply $A \cdot B$ looking at dim(A) 2×3 and dim(B) 3×4 tells us not only that the multiplication is possible, but also that the product matrix will have dimension 2×4 as shown in Figure 13.7.

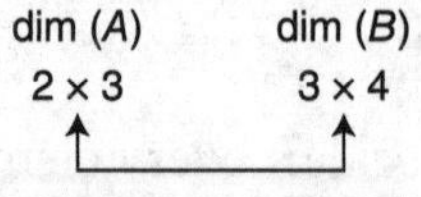

dim (*A*)   dim (*B*)

2×3 3×4

Figure 13.7 These numbers tell the dimension of the product matrix.

Multiplying a single row by a single column

To understand the process of matrix multiplication, we will focus first on a row matrix times a column matrix. Form a row matrix with the prices of the T-shirts and the sweatshirts in our earlier example, [12.99   15.99], and a column matrix with the total sales by the senior class, $\begin{bmatrix} 172 \\ 95 \end{bmatrix}$.

The element 172 represents the number of T-shirts sold by the senior class, and the element 95 is the number of sweatshirts they sold.

In order to find the total amount that the senior class raised, we need to multiply $12.99 times 172 and multiply $15.99 times 95 and add the results together. The first element in the row is multiplied by the first element in the column, and then the second element in the row is multiplied by the second element in the column. These products are added to form the single element in the product matrix. The result of multiplying a 1×2 row matrix times a 2×1 column matrix is a 1×1 matrix.

$$[12.99 \quad 15.99] \cdot \begin{bmatrix} 172 \\ 95 \end{bmatrix} = [12.99 \cdot 172 + 15.99 \cdot 95] = [2234.28 + 1519.05] = [3753.33]$$

If there are more elements in the row and the column—remember that the number of elements in the row must match the number of elements in the column—then there are additional products, but all are combined to produce a single element in the product matrix.

$$[3 \quad 2 \quad 5 \quad 1] \cdot \begin{bmatrix} -2 \\ 1 \\ 4 \\ -3 \end{bmatrix} = [3(-2) + 2(1) + 5(4) + 1(-3)] = [-6 + 2 + 20 + -3] = [13]$$

Multiplying a single row by a larger matrix

In our row-times-column example, we found the total amount of money raised by the senior class. We could repeat the exercise for each of the classes, but it would be simpler if we could multiply the row matrix containing prices [12.99   15.99] by the 2×4 matrix containing the numbers of T-shirts and sweatshirts sold by each class. Multiplying this 1×2 matrix by a 2×4 matrix should produce a 1×4 matrix, which logically would contain the fundraising totals for each of the four classes.

To multiply a row matrix by a matrix with more than one column, multiply the row matrix times the first column of the larger matrix to produce the first element of the product matrix. Repeat the multiplication using the row and each successive column to fill the matrix.

$$[12.99 \quad 15.99] \cdot \begin{bmatrix} 172 & 88 & 106 & 42 \\ 95 & 61 & 75 & 12 \end{bmatrix}$$

$$= [12.99 \times 172 + 15.99 \times 95 \quad ? \quad ? \quad ?]$$

$$= [3753.33 \quad 12.99 \times 88 + 15.99 \times 61 \quad ? \quad ?]$$

$$= [3753.33 \quad 2118.51 \quad 12.99 \times 106 + 15.99 \times 75 \quad ?]$$

$$= [3753.33 \quad 2118.51 \quad 2576.19 \quad 12.99 \times 42 + 15.99 \times 12]$$

$$= [3753.33 \quad 2118.51 \quad 2576.19 \quad 737.46]$$

The multiplication tells us that the senior class raised $3763.33, the junior class $2118.51, the sophomore class $2576.19, and the freshman class $737.46.

Multiplying matrices

The process of multiplying two larger matrices repeats these same steps, with each row of the first matrix producing a row of the product matrix. To multiply two matrices, begin by multiplying the first row of the first matrix by each column of the second matrix, placing the results in the first row of the product matrix. Repeat the process using each row of the first matrix, and place the results in the corresponding row of the product matrix.

As an example, we will multiply the 4×2 matrix $\begin{bmatrix} 1 & 3 \\ 2 & 4 \\ 5 & -1 \\ -2 & -3 \end{bmatrix}$ by the 2×3 matrix $\begin{bmatrix} -2 & 1 & 0 \\ 3 & -1 & 4 \end{bmatrix}$.

A look at the dimensions of the matrices tells us that the multiplication is possible and that the product matrix will be 4×3 (see Figure 13.8).

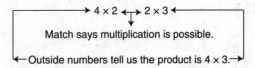

Figure 13.8 Use dimensions to anticipate the product.

$$\begin{bmatrix} 1 & 3 \\ 2 & 4 \\ 5 & -1 \\ -2 & -3 \end{bmatrix} \cdot \begin{bmatrix} -2 & 1 & 0 \\ 3 & -1 & 4 \end{bmatrix} = \begin{bmatrix} ? & ? & ? \\ ? & ? & ? \\ ? & ? & ? \\ ? & ? & ? \end{bmatrix}$$

Focus on the first row.

$$\begin{bmatrix} 1 & 3 \\ 2 & 4 \\ 5 & -1 \\ -2 & -3 \end{bmatrix} \cdot \begin{bmatrix} -2 & 1 & 0 \\ 3 & -1 & 4 \end{bmatrix} = \begin{bmatrix} 1\times(-2)+3\times3 & 1\times1+3\times(-1) & 1\times0+3\times4 \\ ? & ? & ? \\ ? & ? & ? \\ ? & ? & ? \end{bmatrix} = \begin{bmatrix} 7 & -2 & 12 \\ ? & ? & ? \\ ? & ? & ? \\ ? & ? & ? \end{bmatrix}$$

Repeat for the second row.

$$\begin{bmatrix} 1 & 3 \\ 2 & 4 \\ 5 & -1 \\ -2 & -3 \end{bmatrix} \cdot \begin{bmatrix} -2 & 1 & 0 \\ 3 & -1 & 4 \end{bmatrix} = \begin{bmatrix} 7 & -2 & 12 \\ 2\times(-2)+4\times3 & 2\times1+4\times(-1) & 2\times0+4\times4 \\ ? & ? & ? \\ ? & ? & ? \end{bmatrix} = \begin{bmatrix} 7 & -2 & 12 \\ 8 & -2 & 16 \\ ? & ? & ? \\ ? & ? & ? \end{bmatrix}$$

Then the third row.

$$\begin{bmatrix} 1 & 3 \\ 2 & 4 \\ 5 & -1 \\ -2 & -3 \end{bmatrix} \cdot \begin{bmatrix} -2 & 1 & 0 \\ 3 & -1 & 4 \end{bmatrix} = \begin{bmatrix} 7 & -2 & 12 \\ 8 & -2 & 16 \\ 5\times(-2)+(-1)\times3 & 5\times1+-1\times(-1) & 5\times0+(-1)\times4 \\ ? & ? & ? \end{bmatrix} = \begin{bmatrix} 7 & -2 & 12 \\ 8 & -2 & 16 \\ -13 & 6 & -4 \\ ? & ? & ? \end{bmatrix}$$

And finally for the bottom row.

$$\begin{bmatrix} 1 & 3 \\ 2 & 4 \\ 5 & -1 \\ -2 & -3 \end{bmatrix} \cdot \begin{bmatrix} -2 & 1 & 0 \\ 3 & -1 & 4 \end{bmatrix} = \begin{bmatrix} 7 & -2 & 12 \\ 8 & -2 & 16 \\ -13 & 6 & -4 \\ -2\times(-2)+(-3)\times3 & -2\times1+(-3)\times(-1) & -2\times0+(-3)\times4 \end{bmatrix} = \begin{bmatrix} 7 & -2 & 12 \\ 8 & -2 & 16 \\ -13 & 6 & -4 \\ -5 & 1 & -12 \end{bmatrix}$$

The identity matrix

If you multiply $\begin{bmatrix} 1 & 0 \\ 0 & 1 \end{bmatrix}\begin{bmatrix} 2 & 7 \\ 9 & 6 \end{bmatrix}$, you may note an interesting result. The product is identical to the second matrix. $\begin{bmatrix} 1 & 0 \\ 0 & 1 \end{bmatrix}\begin{bmatrix} 2 & 7 \\ 9 & 6 \end{bmatrix} = \begin{bmatrix} 2 & 7 \\ 9 & 6 \end{bmatrix}$. If you explore a little, you'll find that anytime you multiply by a square matrix with 1s on the diagonal and 0s everywhere else, it leaves the other matrix unchanged. It's the matrix equivalent of multiplying a number by 1. A square matrix with 1s on the main diagonal and 0s elsewhere is called an identity matrix.

Identity matrices come in various sizes, but they're always square. To multiply the matrix $\begin{bmatrix} 8 & 4 & 9 & 2 \\ 6 & 1 & 3 & 5 \end{bmatrix}$ by an identity on the left, you'll need a 2×2 identity.

$$\begin{bmatrix} 1 & 0 \\ 0 & 1 \end{bmatrix}\begin{bmatrix} 8 & 4 & 9 & 2 \\ 6 & 1 & 3 & 5 \end{bmatrix} = \begin{bmatrix} 8 & 4 & 9 & 2 \\ 6 & 1 & 3 & 5 \end{bmatrix}$$

To multiply the same matrix $\begin{bmatrix} 8 & 4 & 9 & 2 \\ 6 & 1 & 3 & 5 \end{bmatrix}$ by an identity on the right, you'll need a 4×4 identity.

$$\begin{bmatrix} 8 & 4 & 9 & 2 \\ 6 & 1 & 3 & 5 \end{bmatrix}\begin{bmatrix} 1 & 0 & 0 & 0 \\ 0 & 1 & 0 & 0 \\ 0 & 0 & 1 & 0 \\ 0 & 0 & 0 & 1 \end{bmatrix} = \begin{bmatrix} 8 & 4 & 9 & 2 \\ 6 & 1 & 3 & 5 \end{bmatrix}$$

EXERCISE

13·4

For probs. 1 through 13, use the matrices $A = \begin{bmatrix} 2 & 5 \\ 7 & 3 \end{bmatrix}$, $B = \begin{bmatrix} 5 & 2 & 6 \\ 1 & 0 & 9 \end{bmatrix}$, $C = \begin{bmatrix} 1 & 0 \\ 5 & 2 \\ 7 & 6 \end{bmatrix}$,

$D = [-1 \quad 7 \quad 3]$, *and* $E = \begin{bmatrix} 2 \\ 9 \\ 6 \end{bmatrix}$.

Determine whether each multiplication is possible. If it is possible, give the dimension of the product matrix.

1. $A \times B$

2. $B \times A$

3. $A \times C$

4. $C \times A$

5. $B \times C$

6. $C \times B$

7. $B \times D$

8. $D \times B$

9. $B \times E$

10. $E \times B$

11. $D \times E$

12. $E \times D$

13. $A \times D$

Multiply, if possible.

14. $[1 \quad 3] \begin{bmatrix} 4 \\ 2 \end{bmatrix}$

17. $[5 \quad -2 \quad 1] \begin{bmatrix} 7 & 4 \\ -2 & 6 \\ 9 & 1 \end{bmatrix}$

15. $[3 \quad -2 \quad 1] \begin{bmatrix} 2 \\ 6 \\ 1 \end{bmatrix}$

18. $\begin{bmatrix} 1 & 3 \\ -4 & 2 \end{bmatrix} \begin{bmatrix} 1 & 0 & 5 \\ 3 & 2 & -1 \end{bmatrix}$

16. $[1 \quad 2 \quad 4] \begin{bmatrix} 5 & 4 & 7 \\ 3 & 5 & 8 \end{bmatrix}$

Determinants

The determinant of a matrix is a single number associated with the matrix. Although that is a rather uninformative definition, it is difficult to give a better one. In spite of this difficulty, determinants are important in our study of matrix algebra.

Only square matrices have determinants, and in the simplest case, the determinant of a 1×1 matrix is the single element of the matrix. The determinant of a matrix A is indicated as $|A|$. The bars that indicate the determinant may remind you of the symbols for absolute value, but the significance is quite different. We can use the bars around the entire array rather than the name.

The determinant of the matrix $\begin{bmatrix} 4 & 8 \\ 3 & 1 \end{bmatrix}$ is denoted by $\begin{vmatrix} 4 & 8 \\ 3 & 1 \end{vmatrix}$.

Finding the determinant of a 2 × 2 matrix

In a square matrix, the diagonal path from upper left to lower right is called the major diagonal.

The diagonal from upper right to lower left is the minor diagonal. In the matrix $\begin{bmatrix} 4 & 8 \\ 3 & 1 \end{bmatrix}$, the major diagonal contains the elements 4 and 1, while the minor diagonal contains 8 and 3.

The determinant of a 2×2 matrix is equal to the product of the elements on the major diagonal minus the product of the elements on the minor diagonal. The determinant $\begin{vmatrix} 4 & 8 \\ 3 & 1 \end{vmatrix} = 4 \times 1 - 8 \times 3 = 4 - 24 = -20$. If we write this in symbolic terms, we can say $\begin{vmatrix} a & b \\ c & d \end{vmatrix} = ad - bc$.

The determinant of the square matrix $M = \begin{bmatrix} -2 & 3 \\ 1 & -4 \end{bmatrix}$ can be found quickly by applying the rule:

$|M| = \begin{vmatrix} -2 & 3 \\ 1 & -4 \end{vmatrix} = (-2) \times (-4) - 3 \times 1 = 8 - 3 = 5$. There are several methods for finding the determinants of larger matrices, but considering how tedious the calculations can become, using a calculator or computer is the most popular.

Find each determinant, if possible.

1. $\begin{vmatrix} 6 & 5 \\ 2 & 1 \end{vmatrix}$

2. $\begin{vmatrix} -3 & 7 \\ 4 & -2 \end{vmatrix}$

3. $\begin{vmatrix} 1 & 5 & 3 \\ 8 & 2 & 4 \end{vmatrix}$

4. $\begin{vmatrix} 4 & 7 \\ -5 & 9 \end{vmatrix}$

5. $\begin{vmatrix} 3 & 2 & 5 & 2 \end{vmatrix}$

6. $\begin{vmatrix} 1 & 0 \\ 0 & 1 \end{vmatrix}$

7. $\begin{vmatrix} 7 \end{vmatrix}$

8. $\begin{vmatrix} 5 & 3 \\ 10 & 6 \end{vmatrix}$

Find the missing element.

9. $\begin{vmatrix} ? & 5 \\ 1 & 4 \end{vmatrix} = 7$

10. $\begin{vmatrix} 9 & ? \\ 5 & 3 \end{vmatrix} = 2$

11. $\begin{vmatrix} 4 & 3 \\ ? & 5 \end{vmatrix} = 20$

Inverses

You may have noticed in our discussion of matrix arithmetic that no mention was made of division. While there is no operation of matrix division, the arithmetic of matrices does include the concept of a multiplicative inverse. The idea of the inverse is familiar from standard arithmetic, even if that term is not used as commonly.

In arithmetic, you learned that every non-0 number has a reciprocal and that the product of the number and its reciprocal is 1. The reciprocal of $\frac{3}{5}$, for example, is $\frac{5}{3}$, and the product $\frac{3}{5} \cdot \frac{5}{3} = 1$.

What we commonly refer to as the reciprocal is formally called the multiplicative inverse. Two numbers are multiplicative inverses if their product is 1, the identity element for multiplication.

To transfer the concept of multiplicative inverse to matrix arithmetic, we need to establish a few requirements. In an earlier exercise, we saw that the identity element for matrix multiplication is a square matrix composed of 1s on the major diagonal and 0s elsewhere.

$$\begin{bmatrix} 1 & 0 \\ 0 & 1 \end{bmatrix} \cdot \begin{bmatrix} 3 & 6 & 1 & 8 & -2 \\ 4 & -7 & 0 & 2 & -1 \end{bmatrix} = \begin{bmatrix} 3 & 6 & 1 & 8 & -2 \\ 4 & -7 & 0 & 2 & -1 \end{bmatrix}$$

In order for two matrices to be called inverses, their product must be such an identity matrix. We know, too, that matrix multiplication is not generally commutative, so we add the condition that to be called inverses the two matrices must produce the same identity when multiplied in either order. That requires that the two matrices both be square. If they were not square, then the product $A \times B$ will be of a different size than the product $B \times A$, even if both are identities.

$$\begin{bmatrix} 1 & 2 & 3 \\ 0 & 0 & 1 \end{bmatrix} \cdot \begin{bmatrix} 1 & -1 \\ 0 & -1 \\ 0 & 1 \end{bmatrix} = \begin{bmatrix} 1 & 0 \\ 0 & 1 \end{bmatrix}$$

$$\begin{bmatrix} 1 & -1 \\ 0 & -1 \\ 0 & 1 \end{bmatrix} \cdot \begin{bmatrix} 1 & 2 & 3 \\ 0 & 0 & -1 \end{bmatrix} = \begin{bmatrix} 1 & 2 & 4 \\ 0 & 0 & 1 \\ 0 & 0 & -1 \end{bmatrix}$$

If A and B are square matrices and I is an identity matrix of the same dimension and if $A \times B = B \times A = I$, then A and B are inverse matrices. We denote the inverse of matrix M as M^{-1}.

When considering whether the inverse of a particular matrix exists, it is wise to first calculate the determinant of the matrix in question. Since only square matrices have determinants, this serves as a reminder that only square matrices can have inverses. Not all square matrices actually do have inverses, however, and for reasons we will see in a few moments, those matrices that have determinants of 0 have no inverse. We say that such a matrix is not invertible.

Verifying inverses

To determine whether two matrices are inverses, we must check both possible products. To determine if $A = \begin{bmatrix} 4 & 3 \\ 3 & 2 \end{bmatrix}$ and $B = \begin{bmatrix} -2 & 3 \\ 3 & -4 \end{bmatrix}$ are inverses, we check both the products $A \times B$ and $B \times A$.

$$A \times B = \begin{bmatrix} 4 & 3 \\ 3 & 2 \end{bmatrix} \cdot \begin{bmatrix} -2 & 3 \\ 3 & -4 \end{bmatrix} = \begin{bmatrix} -8+9 & 12-12 \\ -6+6 & 9-8 \end{bmatrix} = \begin{bmatrix} 1 & 0 \\ 0 & 1 \end{bmatrix}$$

$$B \times A = \begin{bmatrix} -2 & 3 \\ 3 & -4 \end{bmatrix} \cdot \begin{bmatrix} 4 & 3 \\ 3 & 2 \end{bmatrix} = \begin{bmatrix} -8+9 & -6+6 \\ 12-12 & 9-8 \end{bmatrix} = \begin{bmatrix} 1 & 0 \\ 0 & 1 \end{bmatrix}$$

Finding the inverse of a 2 × 2 matrix

Given two matrices, verifying whether they are inverses is a simple matter of multiplication. Most often, however, we have only one matrix and need to find its inverse if one exists. Finding the inverse of a 2×2 matrix is a relatively simple process, but for larger matrices, the process becomes more complex.

To find the inverse of a 2×2 matrix,

- Find the determinant of the matrix
- Exchange the elements on the major diagonal
- Change the signs of the elements on the minor diagonal
- Multiply by the reciprocal of the determinant

To find the inverse of the matrix $\begin{bmatrix} 1 & 4 \\ 0 & 2 \end{bmatrix}$, we first find the determinant of the matrix.

$$\begin{vmatrix} 1 & 4 \\ 0 & 2 \end{vmatrix} = 2 - 0 = 2$$

Remember that matrices with determinants of 0 have no inverse. Next the elements on the major diagonal, 1 and 2, exchange places, and the signs of the elements on the minor diagonal are changed. Of course, since 0 is neither positive nor negative, it remains 0.

$$\begin{bmatrix} 1 & 4 \\ 0 & 2 \end{bmatrix} \text{ becomes } \begin{bmatrix} 2 & -4 \\ 0 & 1 \end{bmatrix}.$$

Finally, we perform a scalar multiplication, multiplying by the reciprocal of the determinant, $\frac{1}{2}$. In cases when the determinant is 0, it is impossible to find a reciprocal and the process is stopped.

$$\frac{1}{2} \cdot \begin{bmatrix} 2 & -4 \\ 0 & 1 \end{bmatrix} = \begin{bmatrix} 1 & 2 \\ 0 & \frac{1}{2} \end{bmatrix}$$

The inverse of the matrix $\begin{bmatrix} 1 & 4 \\ 0 & 2 \end{bmatrix}$ is the matrix $\begin{bmatrix} 1 & 2 \\ 0 & \frac{1}{2} \end{bmatrix}$.

EXERCISE

13·6

Determine whether the given matrices are inverses.

1. $\begin{bmatrix} 1 & 3 \\ 2 & 5 \end{bmatrix}$ and $\begin{bmatrix} -5 & 3 \\ 2 & -1 \end{bmatrix}$

2. $\begin{bmatrix} 1 & 3 \\ -2 & 5 \end{bmatrix}$ and $\begin{bmatrix} 5 & -3 \\ 2 & 1 \end{bmatrix}$

3. $\begin{bmatrix} 4 & 5 \\ 9 & 7 \end{bmatrix}$ and $\begin{bmatrix} \frac{1}{4} & \frac{1}{5} \\ \frac{1}{9} & \frac{1}{7} \end{bmatrix}$

Tell whether each matrix has an inverse. Do not find the inverse matrix.

4. $\begin{bmatrix} 5 & 3 \\ -2 & 1 \end{bmatrix}$

7. $\begin{bmatrix} 5 & 2 \\ 10 & 4 \end{bmatrix}$

5. $\begin{bmatrix} 4 & 2 \\ 7 & 1 \\ -3 & 2 \end{bmatrix}$

8. $\begin{bmatrix} 1 & 9 \\ 3 & 2 \end{bmatrix}$

6. $\begin{bmatrix} 9 & 1 \\ 5 & 2 \end{bmatrix}$

Find the inverse of each matrix, if possible.

9. $\begin{bmatrix} 5 & 3 \\ -2 & 1 \end{bmatrix}$

10. $\begin{bmatrix} 9 & 1 \\ 5 & 2 \end{bmatrix}$

Solving systems with matrices

Solving a system of equations—two or more equations in two or more variables—is a common task in algebra. In order to arrive at a solution, you need to have as many equations as variables. There are several methods of solving a system of equations that make use of matrix algebra.

Cramer's rule

Cramer's rule is a method for determining the solution of a system of equations by means of determinants. You probably remember from algebra the elimination method of solving a system of equations. One or both equations can be multiplied by a constant and then the equations added to eliminate one of the variables. Solving the system $\begin{cases} 2x + 3y = 19 \\ 3x - 4y = 3 \end{cases}$ will allow us to look at the options available. If we choose to eliminate x from the equations, the first equation can be

multiplied by 3 and the second equation multiplied by –2.
Adding the equations eliminates x.

$$6x + 9y = 57$$
$$\underline{-6x + 8y = -6}$$
$$17y = 51$$
$$y = \frac{51}{17} = 3$$

If instead we choose to eliminate y, the first equation is multiplied by 4 and the second by 3.

Then the equations are added.

$$8x + 12y = 76$$
$$9x - 12y = 9$$
$$\overline{17x \qquad = 85}$$

$$x = \frac{85}{17} = 5$$

To understand the origins of Cramer's rule, we want to focus for a moment on two lines, one from each of the solutions above: $x = \frac{85}{17} = 5$ and $y = \frac{51}{17} = 3$. The denominator of 17, common to both, is equal to $2 \cdot 4 + 3 \cdot 3$, the opposite of the determinant of the matrix of the coefficients of x and y: $\begin{vmatrix} 2 & 3 \\ 3 & -4 \end{vmatrix} = 2(-4) - 3(3) = -8 - 9 = -17$. The 85 can be produced from a determinant involving the coefficients of the y-terms and the constants, and the 51 from the coefficients of the x-terms and the constants. Cramer's rule recognizes this and uses determinants to arrive at the solution of the system quickly and easily.

The denominator

To use Cramer's rule to solve a system, we first find the determinant of the matrix of coefficients, placing the x coefficients in the first column and the y coefficients in the second: $\begin{vmatrix} 2 & 3 \\ 3 & -4 \end{vmatrix} = 2(-4) - 3(3) = -8 - 9 = -17$. This determinant will be our denominator for both x and y.

A numerator for each variable

Next we create a determinant for each variable. The numerator for x is the determinant formed when we take the coefficient matrix $\begin{vmatrix} 2 & 3 \\ 3 & -4 \end{vmatrix}$ and replace the values in the x column with the constants.

$$\begin{vmatrix} 19 & 3 \\ 3 & -4 \end{vmatrix} = 19(-4) - 3(3) = -76 - 9 = -85$$

The numerator for y is the determinant formed when we take the coefficient matrix $\begin{vmatrix} 2 & 3 \\ 3 & -4 \end{vmatrix}$ and replace the values in the y column with the constants.

$$\begin{vmatrix} 2 & 19 \\ 3 & 3 \end{vmatrix} = 2(3) - 3(19) = 6 - 57 = -51$$

Notice that both of these have come out opposite in sign to the values we saw in the algebraic solutions. In a moment, when we divide, these sign differences will cancel one another.

Solutions

To find the values of x and y, all that remains is to divide and simplify. The value of x is the x numerator over the denominator: $x = \frac{-85}{-17} = 5$. The value of y is the y numerator over the denominator: $y = \frac{-51}{-17} = 3$.

Solve each system by Cramer's rule.

1.
$$-3x+2y=-20$$
$$5x-3y=33$$

3.
$$2x+3y=2.9$$
$$4x-y=-3.3$$

2.
$$2x-7y=1$$
$$3x+5y=17$$

Method of inverses

We have already seen that Cramer's rule allows us to solve systems of equations by means of the determinants of matrices formed from the coefficients of the system. With the addition of inverses to the matrix tool kit, we can expand our understanding of matrix algebra to access an additional method.

From your experience solving systems, you no doubt recognize that there is no difference in the solution of the system $\begin{cases} 3x-2y=13 \\ x+3y=8 \end{cases}$ and the system $\begin{cases} 3a-2b=13 \\ a+3b=8 \end{cases}$. You know that the particular letters used as variables are less significant than the coefficients and constants that define the system. All the matrix methods of solving systems depend on matrices made from these coefficients and constants, but unlike Cramer's rule, these methods do not replace the coefficients with the constants.

Matrices of coefficients and of constants

If we use the system $\begin{cases} 3x-2y=13 \\ x+3y=8 \end{cases}$ from our earlier discussion as our example, we can form two matrices. The first is a 2×2 matrix containing the coefficients of x in the first column and the coefficients of y in the second column. Each equation forms a row of the matrix, so the matrix of coefficients is $\begin{bmatrix} 3 & -2 \\ 1 & 3 \end{bmatrix}$. The second matrix, a column matrix with dimension 2×1, contains the constants from the other side of each equation. The matrix of constants thus formed is $\begin{bmatrix} 13 \\ 8 \end{bmatrix}$.

The key to this method lies in the fact that the system can be represented by a single statement of matrix multiplication. If the coefficient matrix is multiplied by a column matrix containing our variables, the result is exactly the left side of our system.

$$\begin{bmatrix} 3 & -2 \\ 1 & 3 \end{bmatrix} \cdot \begin{bmatrix} x \\ y \end{bmatrix} = \begin{bmatrix} 3x-2y \\ x+3y \end{bmatrix}$$

Since this is the case, our system of equations is equivalent to the matrix equation

$$\begin{bmatrix} 3 & -2 \\ 1 & 3 \end{bmatrix} \cdot \begin{bmatrix} x \\ y \end{bmatrix} = \begin{bmatrix} 13 \\ 8 \end{bmatrix}$$

The algebra of matrices allows us to multiply both sides of this equation by the same matrix, just as standard algebra allows us to multiply both sides of an equation by the same number. If then we choose to multiply both sides of this equation by the inverse of the coefficient matrix, we find ourselves with a rapid solution.

$$\underbrace{\begin{bmatrix} 3 & -2 \\ 1 & 3 \end{bmatrix}^{-1} \cdot \begin{bmatrix} 3 & -2 \\ 1 & 3 \end{bmatrix}}_{\begin{bmatrix} 1 & 0 \\ 0 & 1 \end{bmatrix}} \cdot \begin{bmatrix} x \\ y \end{bmatrix} = \begin{bmatrix} 3 & -2 \\ 1 & 3 \end{bmatrix}^{-1} \cdot \begin{bmatrix} 13 \\ 8 \end{bmatrix}$$

$$\begin{bmatrix} 1 & 0 \\ 0 & 1 \end{bmatrix}\begin{bmatrix} x \\ y \end{bmatrix} = \begin{bmatrix} 3 & -2 \\ 1 & 3 \end{bmatrix}^{-1} \cdot \begin{bmatrix} 13 \\ 8 \end{bmatrix}$$

$$\begin{bmatrix} x \\ y \end{bmatrix} = \begin{bmatrix} 3 & -2 \\ 1 & 3 \end{bmatrix}^{-1} \cdot \begin{bmatrix} 13 \\ 8 \end{bmatrix}$$

The solution of the system of equations is equal to the inverse of the coefficient matrix times the constant matrix.

In our 2×2 example, $\begin{bmatrix} x \\ y \end{bmatrix} = \begin{bmatrix} 3 & -2 \\ 1 & 3 \end{bmatrix}^{-1} \cdot \begin{bmatrix} 13 \\ 8 \end{bmatrix}$. Since $\begin{bmatrix} 3 & -2 \\ 1 & 3 \end{bmatrix}^{-1} = \frac{1}{11}\begin{bmatrix} 3 & 2 \\ -1 & 3 \end{bmatrix} = \begin{bmatrix} \frac{3}{11} & \frac{2}{11} \\ \frac{-1}{11} & \frac{3}{11} \end{bmatrix}$, we can

find our solution by multiplying.

$$\begin{bmatrix} x \\ y \end{bmatrix} = \begin{bmatrix} \frac{3}{11} & \frac{2}{11} \\ \frac{-1}{11} & \frac{3}{11} \end{bmatrix} \cdot \begin{bmatrix} 13 \\ 8 \end{bmatrix} = \begin{bmatrix} \frac{39}{11}+\frac{16}{11} \\ \frac{-13}{11}+\frac{24}{11} \end{bmatrix} = \begin{bmatrix} \frac{55}{11} \\ \frac{11}{11} \end{bmatrix} = \begin{bmatrix} 5 \\ 1 \end{bmatrix}$$

Thus, we know that $x = 5$ and $y = 1$.

Although we have used a system of two equations in two variables as our example, the logic of the matrix algebra holds for systems of any size. Since calculators allow us to find inverses and to multiply matrices easily, the technique gives us an easy way to solve systems of any size.

EXERCISE 13·8

Use inverse matrices to solve each system.

1. $-3x+2y=-20$
   $5x-3y=33$

2. $2x-7y=1$
   $3x+5y=17$

3. $2x+3y=2.9$
   $4x-y=-3.3$

4. $2x+7y=13$
   $x-5y=-2$

5. $3x-5y=36$
   $2x+y=11$

6. $5x+2y=-21$
   $3x-4y=-23$

Calculator notes #10

Not every algebra course includes a look at matrices; in fact, the algebra of matrices can be a course in its own right. If you are familiar with matrices, however, they can provide a powerful way to solve systems of linear equations, especially with the help of your graphing calculator.

To enter a matrix into your calculator:

- Press [2nd] [x⁻¹] to access the MATRIX menu.
- Cursor to EDIT and press [ENTER].
- Choose a slot for your matrix and press [ENTER].

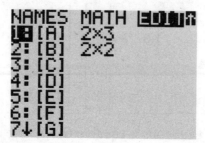

- Enter the dimensions of your matrix: number of rows, [ENTER], number of columns, [ENTER]. The display will adjust to that size.

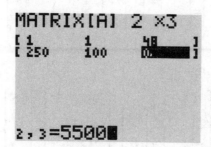

- Enter your matrix, moving across the row, pressing [ENTER] after each number.
- When you have entered the whole matrix, press [2nd] [MODE] to QUIT.

To perform addition, subtraction, or multiplication with matrices you have entered:

- Press [2nd] [x⁻¹] to access the MATRIX menu.
- Choose a matrix and press [ENTER].
- Press the operation sign.
- Press [2nd] [x⁻¹] to access the MATRIX menu.
- Choose the second matrix and press [ENTER].
- Press [ENTER] again to display the result.

The algebra of matrices explains, among other things, how to solve a matrix equation by matrix operations. If A is a square matrix of coefficients, X is a column matrix of variables, and B is a column matrix of constants, the matrix equation $AX = B$ can be solved by multiplying both sides by the inverse of matrix A.

$$AX = B$$

$$\left(A^{-1}A\right)X = A^{-1}B$$

$$X = A^{-1}B$$

With the help of a graphing calculator, this is a convenient way to solve a system of linear equations.

$$5x - 3y = 1$$

$$2x + 5y = 19$$

The system of equations shown above can be expressed as the matrix equation:

$$\begin{bmatrix} 5 & -3 \\ 2 & 5 \end{bmatrix} \begin{bmatrix} x \\ y \end{bmatrix} = \begin{bmatrix} 1 \\ 19 \end{bmatrix}$$

- Enter $\begin{bmatrix} 5 & -3 \\ 2 & 5 \end{bmatrix}$ as matrix A.

- Enter $\begin{bmatrix} 1 \\ 19 \end{bmatrix}$ as matrix B.

- Press ⎡2nd⎤ ⎡x⁻¹⎤ to access the MATRIX menu, choose matrix A, press ⎡x⁻¹⎤, press ⎡2nd⎤ ⎡x⁻¹⎤, choose matrix B, and press ⎡ENTER⎤.
- The solution will be displayed as a column matrix. The first row is the value of x, and the second row is the value of y.

Problem Solving

The branch of mathematics labeled algebra includes many different skills: evaluating expressions, factoring, operating with polynomials or rational expressions, solving various types of equations, and more. Those skills are ultimately tools to solve problems, and yet sometimes the appearance of a "word problem" or "story problem" feels like a new challenge. In this chapter, you'll find some general suggestions for problem solving as well as a look at some of the categories of problems you'll encounter most often.

General strategies

If you were asked to solve the equation $3x - 7 = 14$, you probably wouldn't have any hesitation. You have the skills necessary to do that. You may not jump into solving the problem below as quickly, but in fact, it's the same task.

> Keisha made and sold 3 bracelets at a craft fair. After she paid $7 for her materials, she had a profit of $14. How much did she charge for 1 bracelet?

The first thing to remember when you face a problem is that you have the skills. The problem is not a new adventure. It's just a different way of asking you to do things you know how to do.

The second point seems obvious but it's often forgotten. Look at this problem.

Now, that probably seems silly, but too often that's how a problem looks to people. They grab the numbers—1, 3, 7, and 14—and then try to figure out what to do with them. It's not usually a successful strategy. There's a lot of information going to waste. So the second thing to remember is to take the time to read the problem. You're not expected to magically know what to do with the numbers. Read the problem to find that out.

Your job is to translate the problem into an equation, so think about what you need for an equation. There wouldn't be a need to solve unless there's a variable, some unknown. Locate that first. (Hint: It's often in the last sentence.) Choose a variable and write down what it stands for.

> Keisha made and sold 3 bracelets at a craft fair. After she paid $7 for her materials, she had a profit of $14. *How much did she charge for 1 bracelet?*

Let x = the cost of 1 bracelet

Pause for a moment to think about units. This one is not too difficult. The cost of a bracelet is probably in dollars (or dollars and cents). Measurements could be in feet or inches or meters or centimeters or something else. What units are you asked to use? Is all the information in compatible units? If you have to convert some quantity to different units, make a choice of what you'll convert. Don't convert several if you can make do with converting one. If you have a choice, think about which conversion is easiest.

Just as it doesn't make sense to skip over all the words, it's also unwise to get lost in too many words. Focus on what you really need to know.

Keisha made and *sold 3 bracelets* at a craft fair. After *she paid $7* for her materials, she had a *profit of $14*. How much did she charge for 1 bracelet?

Try to create one sentence that sums up the important information.

3 bracelets at $x each minus $7 leaves $14

Translate that into algebraic symbols.

(3 bracelets at $x each) minus ($7) leaves ($14)

$$(3 \cdot x) - (7) = (14)$$
$$3x - 7 = 14$$

Once you've gotten to this point, you're in familiar territory. You know how to solve this equation. You have the skills.

But it's not always that easy, you say? There are some types of problems that will require more preparation before you can write the equation and those will be covered later in this chapter. Some will require that you know a common formula, like the area of a triangle or the Pythagorean theorem, but many problems include the needed formula, especially if it is one not used regularly.

The complication you may face is that you know how to solve several different types of equations. In the example above, the equation was a first degree or linear equation. The problem didn't lead to a quadratic equation because there was nothing that referred to squaring, nor any situation in which two variable quantities are multiplied together. That might come up in a problem about area, or one which specifically talks about squaring. The point is that your job is just to translate what it says. You don't have to worry about absolute value, or square roots, or rational expressions, or quadratics, unless the problem talks about them.

Like any translation exercise, you need to know your vocabulary and your grammar. Review Chapter 1, Section 1.5 if you need a refresher. Practice with this exercise before looking at the specific types of problems.

EXERCISE
14·1

For each problem, define a variable and write an equation, but do not solve.

1. If 9 is subtracted from a number, the result is 75. What is the number?

2. If 3 times a certain number is decreased by 17, the result is 43. What is the number?

3. Five times a certain number equals 28 more than that number. What is the number?

4. If 8 times a certain number is decreased by 40, the result is 5 less than the original number. What is the number?

5. A tennis player played 30 games, winning 6 more than she lost. How many games did she win?

6. A father is 5 times as old as his son. The difference between their ages is 44 years. How old is the son?

7. A pen costs 89 cents more than a pencil. Together they cost $1.25. Find the cost of the pen.

8. Carlos and Jaden run on the high school track every evening after work. Combined, they run 6.25 miles but Carlos runs 3 more laps than Jaden. If each lap is $\frac{1}{4}$ mile, how many laps does Jaden run?

9. Isaac bought 2 movie tickets and a large bucket of popcorn for $35. A ticket costs twice as much as a large bucket of popcorn. How much does the large bucket of popcorn cost?

10. Kate spends $5.50 a day to ride the bus to and from work, and once a week she spends $8 to have lunch with a friend. When she got paid, Kate set aside $110 for those expenses. Assuming Kate works 5 days a week, how many weeks will that $110 last?

Problems solved with linear equations

Many problems can be solved by linear equations, but there are certain types of problems that follow a clear pattern. Knowing that routine setup for each type of problem makes it easier to translate the problem into an equation. Here are a few of the common types.

Percentage problems

Problems that involve adding tax or a tip to a bill, or applying a discount to a purchase often require you to use percentages. Remember that a percent is just another way of expressing a part or a fraction of a whole. The whole is 100%. A percent greater than 100% indicates more than the whole.

When you calculate, you convert percents to decimals (or fractions): 100% = 1.00, 40% = 0.40, and 3% = 0.03. If you add a tax of 6% to a bill, you'll have 100% of the bill plus another 6% of the bill, or 106% of the bill. If you receive a 15% discount on your bill, you pay 100% of the bill minus 15% of the bill, or 85% of the bill.

Jeff sees an advertisement for a television at an attractive price. Because Jeff has a loyalty card for the vendor, he receives a 10% discount. When he purchases the TV, the discount is applied, but after that, a 7% tax is added. If Jeff pays $383.27, what was the advertised price to the nearest dollar?

Let x = the advertised price of the TV

Let $0.90x$ = what Jeff paid, before tax

Let $1.07(0.90x)$ = total Jeff paid

The total Jeff paid was $383.27.

$$1.07(0.90x) = \$383.27$$
$$0.963x = 383.27$$
$$x = \frac{383.27}{0.963}$$
$$x \approx 398$$

The advertised price of the television, rounded to the nearest dollar, is $398.

Mixture problems

In Chapter 2, you spent some time looking at the general category of mixture problems. Whether they mix coins of different denominators or different types of nuts, candies, teas, or coffees, or even vehicles traveling at different speeds, they can all be made more manageable by organizing the given information into a chart.

The chart should have a row for each individual item and one for the mixture. The columns may show how much (or how many) of the item, the value of one unit of that item, and the total value. In problems about traveling, the columns are the rate of speed, the time traveled, and the total distance. In each case, adding down the last column produces the equation.

Mixing chemicals instead of coffee? The approach is the same, but the column headings change. Your equation still comes from adding down the last column.

A chemist has a 30% solution of a certain chemical, and a pure version of the chemical, but she needs 100 grams of a 65% solution. How many grams of the 30% solution and how many grams of the pure chemical should she mix to produce 100 grams of a 65% solution?

Let x = the number of grams of the 30% solution

$100 - x$ = the number of grams of pure chemical

	Grams of solution	Percent of chemical	Grams of chemical in this solution
30% solution	x	0.30	$0.3x$
Pure chemical	$100 - x$	1.00	$100 - x$
65% solution	100	0.65	65

$$0.3x + (100 - x) = 65$$
$$-0.7x + 100 = 65$$
$$-0.7x = -35$$
$$x = \frac{-35}{-0.7} = 50$$

The chemist should use 50 grams of the 30% solution and 50 grams of the pure chemical.

Perimeter problems

Perimeter is the word used to mean the distance around the outside edges of a closed figure. The perimeter of a triangle is the sum of the lengths of its three sides. For some figures, it's possible to give a formula for the perimeter. The perimeter of a rectangle is $P = 2l + 2w$, the total of two lengths and two widths. For a square, the perimeter is $P = 4s$ because all sides are the same length. When no particular formula exists, you simply add the lengths of the sides. Perimeter problems will usually tell you how the lengths of sides compare to one another so that you can write an equation that doesn't involve many different variables.

Each of the two equal sides of an isosceles triangle is 4 centimeters less than twice the third side. The perimeter is 52 centimeters. Find the length of each side.

Let x = the length of the non-equal side

Let $2x - 4$ = the length of each of the equal sides

The perimeter of the triangle is 52 centimeters.

$$x + (2x - 4) + (2x - 4) = 52$$
$$5x - 8 = 52$$
$$5x = 60$$
$$x = 12$$
$$2x - 4 = 20$$

The two equal sides are each 20 centimeters and the third side is 12 centimeters.

Consecutive integer problems

Integers are positive or negative whole numbers and zero. Consecutive integers are integers that follow one another in counting order. The integers 7, 8, and 9 are consecutive integers. If you need to find consecutive integers, you can let $x =$ the smallest integer, like the 7 in the example. Then the second integer can be $x + 1$, and the third can be $x + 2$.

Consecutive even integers, like 14, 16, and 18, can be represented as x, $x + 2$, and $x + 4$, because you're counting by 2, not by 1. Consecutive even integers are every other number. Many people are surprised to learn that consecutive odd integers, like 11, 13, and 15, are also represented as x, $x + 2$, and $x + 4$. You're still taking every other number. The difference is just that you're starting with an odd number instead of an even number.

There are five consecutive integers. If the sum of the last four is decreased by the first, the result is 100. Find the five integers.

Let $x =$ the first integer

Let $x + 1 =$ the second integer

Let $x + 2 =$ the third integer

Let $x + 3 =$ the fourth integer

Let $x + 4 =$ the fifth integer

The sum of the last four integers decreased by the first integer is 100.

$$\left[(x+1) + (x+2) + (x+3) + (x+4)\right] - x = 100$$
$$(4x + 10) - x = 100$$
$$3x + 10 = 100$$
$$3x = 90$$
$$x = 30$$

The five consecutive integers are 30, 31, 32, 33, and 34.

Direct variation problems

The ideas of direct variation and combined variation have been covered in Chapter 10. If y varies directly with x, y goes up when x goes up, y goes down when x goes down, and there is a constant of variation, k, that connects x and y. The direct variation equation always has the form $y = kx$. The constant of variation is the slope of the linear equation, and in direct variation, the y-intercept is always zero.

Word problems about direct variation don't sound very different than the exercises you've seen. Instead of "y varies directly with x," the problem will use the names of the quantities x and *y* represent. Use information given in the problem to find *k*, and then use the specific equation for the problem to find the value the problem asks for.

Water pressure, *P*, varies directly with the height, *h*, of the water. When the height is 16 feet, the pressure is 1,000 pounds per square foot. Find the pressure for a height of 10 feet.

Let *P* = the water pressure in pounds per square foot

Let *h* = the height in feet

$$P = kh$$

$$1,000 = k(16)$$

$$\frac{1,000}{16} = k$$

$$k = 62.5$$

The specific equation for this situation is *P* = 62.5*h*.

$$P = 62.5(10) = 625$$

For a height of 10 feet, the pressure is 625 pounds per square foot.

EXERCISE
14·2

1. How many ounces of chocolate that is 60% cocoa should be mixed with 4 ounces of chocolate that is 80% cocoa to produce a mixture that is 75% cocoa?

2. The length of a rectangle is 8 feet less than 3 times the width. The perimeter of the rectangle is 100 feet. Find the length and the width.

3. The area of a rectangle varies directly with its length. When the length is 15 inches, the area is 127.5 square inches. What length will produce an area of 229.5 square inches?

4. If 35% of a certain number is increased by 27, the result will be 80% of the original number. Find the original number.

5. Find 3 consecutive integers whose sum is 96.

6. A vaccine is available in full strength (100%) and 50% solutions, but latest research shows that the safe and effective concentration for children is 65%. How much full-strength vaccine and how much 50% solution should be mixed to produce 100 mg of a 65% solution?

7. The five sides of a pentagon, measured in inches, are consecutive integers. Find the length of each side if the perimeter is 450 inches.

8. Find 4 consecutive even integers whose sum is 124.

9. If your salary is reduced by 10%, by what percent must your new salary be increased to restore your original salary? [Hint: You can make up your own salary, if it helps.]

10. The area of a triangle varies directly with its height. A height of 12.3 cm results in an area of 110.7 cm². What height will produce an area of 108 cm²?

Problems solved with quadratic equations

Many problems can be solved by using a linear equation, but some problems require other types of equations. Quadratic equations come up frequently and easily. A direct variation equation is usually solved with a linear equation of the form $y = kx$ but if the problem were to say "y varies directly with the square of x" the equation becomes $y = kx^2$. That's a simple quadratic equation.

When solving these problems, remember to simplify each side of the equation, and then move all terms to one side, so that you have an equation of the form $ax^2 + bx + c = 0$. Solve by factoring, if possible, or use the quadratic formula.

There are many problems in which that change of language, talking about the square of some variable, leads to a quadratic equation. In other problems, the squaring is not mentioned directly, but arises from a formula or because quantities in the problem need to be multiplied.

Pythagorean theorem problems

One rule you've probably used many times that leads to equations with squares is the Pythagorean theorem. This right triangle relationship that says that if a and b are the lengths of the legs of a right triangle, the length of the hypotenuse, c, can be found using the rule $c^2 = a^2 + b^2$.

The longer leg of a right triangle is 14 inches longer than the shorter leg, and 2 inches shorter than the hypotenuse. Find the lengths of the three sides.

Let $x =$ the length of the shorter leg in inches

Let $x + 14 =$ the length of the longer leg in inches

Let $(x + 14) + 2 = x + 16 =$ the length of the hypotenuse

$$c^2 = a^2 + b^2$$

$$(x + 16)^2 = x^2 + (x + 14)^2$$

$$x^2 + 32x + 256 = x^2 + x^2 + 28x + 196$$

$$x^2 + 32x + 256 = 2x^2 + 28x + 196$$

$$32x + 256 = x^2 + 28x + 196$$

$$256 = x^2 - 4x + 196$$

$$0 = x^2 - 4x - 60$$

$$(x + 6)(x - 10) = 0$$

$$x + 6 = 0 \qquad x - 10 = 0$$

$$x = -6 \qquad x = 10$$

It is not reasonable to have a negative side length, so the shorter leg is 10 inches, the longer leg is 10 + 14 = 24 inches, and the hypotenuse is 24 + 2 = 26 inches.

Area problems

Area problems often give rise to quadratic equations because they require that you multiply length times width or base times height. If both quantities involve the variable, the multiplication results in a variable squared.

If one side of a square is increased by 3 feet and the adjacent side is decreased by 2 feet, a rectangle is created that has an area of 36 square feet. Find the area of the original square.

Let x = the length of a side of the square, in feet

Let $x + 3$ = the length of the longer side of the rectangle

Let $x - 2$ = the length of the shorter side of the rectangle

The area of the rectangle is 36 square feet.

$$(x+3)(x-2)=36$$
$$x^2+x-6=36$$
$$x^2+x-42=0$$
$$(x+7)(x-6)=0$$
$$x+7=0 \qquad x-6=0$$
$$x=-7 \qquad x=6$$

A square cannot have a side with a negative length, so the side of the square is 6 feet, and the area of the square is 36 square feet.

Vertical motion problems

When an object is dropped from a height or thrown upward and allowed to fall back to Earth, its height at any time after it is released depends on three factors: the height from which it was released, the force with which it is thrown, and the force of gravity pulling it back to Earth. This relationship is summarized by the function $h(t) = -\frac{1}{2}gt^2 + vt + s$. In this equation $h(t)$ is the height of the object at time t, g is the acceleration due to gravity (32 feet per second per second in customary units, or 9.8 meters per second per second in metric), and s is the starting height of the object. The v in the middle term refers to the initial velocity of the object. If the object is thrown upward, v is a positive number. If the object is thrown down from a height, v is negative, but if the object is simply released and allowed to fall, $v = 0$. Be sure to check that units of measurement are consistent.

The chart below summarizes the different forms of the function for different circumstances.

	Customary Units	Metric Units
Dropped	$h(t) = -16t^2 + s$	$h(t) = -4.9t^2 + s$
Thrown	$h(t) = -16t^2 + vt + s$	$h(t) = -4.9t^2 + vt + s$

Problems about dropped objects can usually be solved by the square root method, but problems about objects thrown will often require the quadratic formula. Most will have two solutions but only one will make sense in the context of the problem. A negative number for time would refer to before it was thrown, which is not reasonable, so the positive solution would be the one you want. If there are two positive values for time, one indicates the time at which the object reaches that height on the way up, and the other on the way down.

Mr. Patel and his daughter, Deva, like to play catch after dinner. Mr. Patel throws the ball to Deva with an initial velocity of 20 meters per second. When it leaves his hand it is 2 meters above the ground. Deva is most successful in catching the ball when it is 1 meter above the ground. How long after her father throws the ball to Deva will it be 1 meter off the ground?

Let t = the time, in seconds, at which the ball is one meter off the ground

Let $\frac{1}{2}g = \frac{1}{2}(9.8) = 4.9$ meters per second per second

Let $v = 20$ meters per second

Let $s = 2$ meters

The particular equation is $h(t) = -4.9t^2 + 20t + 2$.

Find t when $h(t) = 1$.

Solve $1 = -4.9t^2 + 20t + 2$ with quadratic formula.

$$1 = -4.9t^2 + 20t + 2$$

$$0 = -4.9t^2 + 20t + 1$$

$$a = -4.9 \quad b = 20 \quad c = 1$$

$$x = \frac{-b \pm \sqrt{b^2 - 4ac}}{2a}$$

$$x = \frac{-20 \pm \sqrt{(20)^2 - 4(-4.9)(1)}}{2(-4.9)}$$

$$x = \frac{-20 \pm \sqrt{400 + 19.6}}{-9.8}$$

$$x = \frac{-20 \pm \sqrt{419.6}}{-9.8}$$

$$x = \frac{-20 + \sqrt{419.6}}{-9.8}$$

$$x \approx \frac{0.484}{-9.8} \qquad x \approx \frac{-40.484}{-9.8}$$

$$x \approx -0.049 \qquad x \approx 4.131$$

Deva can't catch the ball before her father throws it, so ignore the negative solution. The ball will be 1 meter off the ground approximately 4.131 seconds after Mr. Patel throws it.

EXERCISE
14·3

1. The square of a number is 85 less than 22 times the number. What is the number?

2. You have a ladder 20 feet long, and want to lean it against a wall so that the top of the ladder reaches a point 17 feet above the floor. How far from the base of the wall should the foot of the ladder be? [Round to the nearest tenth.]

3. Find three consecutive odd integers such that the square of the first is 3 less than the sum of the other two.

4. A triangle has an area of 45 square meters. The height of the triangle is 3 meters more than twice its base. How long is the base of the triangle?

5. The 86th floor observation deck of the Empire State Building is 320 meters above street level. If an object were dropped from the deck, how long would it take to hit the ground?

6. The perimeter of a rectangle is 28 inches. If the diagonal is 10 inches long, find the dimensions of the rectangle.

7. Before a football game, the referee tosses a coin upward with an initial velocity of 7 meters per second. One team captain must call heads or tails before the coin hits the ground. If the referee's hand is 1.8 meters above the ground when he tosses the coin, how long does the team captain have to make a call before the coin hits the ground? [Round to the nearest tenth.]

8. A rectangle is 3 times as long as it is wide. If the width is increased by 6 feet, and the length is decreased by 3 feet, the area is doubled. Find the dimensions of the original rectangle.

9. The sum of two numbers is 5. The sum of the squares of the numbers is 53. Find the numbers.

10. Sanjai walked down the stairs from his apartment to the street and then discovered he had forgotten his wallet. His wife came to the window and offered to toss it out so he didn't have to walk back up. She threw it downward with an initial velocity of 5 meters per second from a height of 13.7 meters. How long did it take to hit the ground? [Round to the nearest tenth.]

Problems solved with rational equations

Rational equations always present a challenge, if only because they can become large and unwieldy. Remember that multiplying through the equation by the simplest common denominator for all rational expressions will clear all the denominators. That may leave a linear equation or a quadratic equation, but solving always seems easier when the denominators are gone.

Extraneous solutions are common in rational equations. Whether you find yourself with one solution or two, remember to check each solution in the original equation. Don't rely on the version after you clear denominators, because you could have had an earlier error.

The types of problems that most often lead to rational equations are problems about shared work and inverse variation problems.

Work problems

Problems that deal with how long it takes two people (or two machines) to do a job while working together can be organized much the same way as all the mixture problems you saw earlier, if you know one little trick. As soon as you are told how long it takes someone to do the job, express the part of the job they can do in one unit of time—one minute, one hour, one day, whatever.

Suppose George can do a job in two hours and Harry can do it in three hours. How long will it take them to do it working together? You want to set up a table similar to the ones for mixture problems.

	Part of the job done in one hour	Number of hours	Whole job
George	$\frac{1}{2}$	2	1
Harry	$\frac{1}{3}$	3	1
Together	$\frac{1}{x}$	x	1

The part of the job done in one hour, times the number of hours, will always equal one. Your equation comes not from adding the hours, but from adding the part of the job done in an hour.

$$\frac{1}{2}+\frac{1}{3}=\frac{1}{x}$$

$$6x\left(\frac{1}{2}+\frac{1}{3}\right)=\left(\frac{1}{x}\right)6x$$

$$3x+2x=6$$

$$5x=6$$

$$x=\frac{6}{5}=1\frac{1}{5}$$

It will take $1\frac{1}{5}$ hours or 1 hour and 12 minutes for them to do the job together.

Inverse variation problems

Like direct variation problems, inverse variation problems are not very different from the basic statement of the problem. Instead of y and x, you'll find the names of the quantities that vary inversely. Unlike direct variation, inverse variation problems will have variables in the denominator of a rational expression, so the solution will require solving simple rational equations.

At a fixed temperature, the volume of a certain quantity of gas varies inversely with the pressure. If the volume is 16 in³ when the pressure is 18 psi, find the pressure that produces a volume of 20 in³.

Let V = the volume in cubic inches

Let P = the pressure in psi (pounds per square inch)

Write the general form of the inverse variation equation.
Substitute known values to find the constant of variation.

$$V=\frac{k}{P}$$

$$16=\frac{k}{18}$$

$$k=288$$

Write the specific equation for this situation. Substitute and solve.

$$V=\frac{288}{P}$$

$$20=\frac{288}{P}$$

$$20P=288$$

$$P=14.4$$

A pressure of 14.4 psi will produce a volume of 20 in³.

1. Two machines in the school office can print final exams. The larger machine can print all the necessary exams in 12 hours, but the smaller one would need 36 hours to do the same job. How long will it take to print all the necessary exams if both machines were used?

2. Elena can complete a quilt in six months, but if she and her mother work on it together, they can complete it in two months. How long would it take Elena's mom to complete the quilt by herself?

3. Every year, Mr. Song has to produce an annual report that fills many, many pages. If he types it himself, it takes him a week. If he gives it to his administrative assistant to type, she can do it in three days. How long will it take to get the annual report typed if both of them work on it?

4. If the drain pipe in the pool is opened, the pool will drain in eight hours. If, in addition to opening the drain, a pump is set up to pump water out, the pool will be emptied in 5 hours. How long would it take the pump, working alone, to empty the pool?

5. Alicia takes six hours longer than David to complete a project. Working together, they can complete the project in 7.2 hours. How long would it take Alicia to complete the project alone?

6. It takes Ron twice as long to mow his lawn with a push mower as it takes him to mow it with a power mower. He starts working with the power mower but after 45 minutes, he runs out of gas and has to finish the job with the push mower. It takes him an additional 45 minutes to finish the job with the push mower. How long would it have taken Ron to mow the entire lawn with the power mower?

7. For a fixed area, the length and width of a rectangle are inversely related. When the length is 12 cm, the width is 9 cm. Find the width when the length is 27 cm.

8. The time required to drive a fixed distance varies inversely with speed. If a trip takes 6.5 hours at 55 mph, how long will it take if driven at 65 mph?

9. The force of gravity on an object varies inversely with its distance from the center of the Earth. The radius of the Earth is 6378 km. The International Space Station orbits approximately 350 km above the Earth, and the force of gravity acting on it is approximately 2.7×10^{12} Newtons. What would the force of gravity be if the International Space Station were 500 km above the surface of the Earth?

10. The illumination on a surface varies inversely with the square of its distance from a light source. At a distance of 5 ft from the light source, the illumination is 270 foot-candles. At what distance from the same light source will the illumination be 750 foot-candles?

Problems solved with systems

Many of the problems you solved with linear equations could also be solved by means of a system of equations. If a mixture problem talks about making 10 pounds of a tea blend, you could let x represent the amount of one tea and $10 - x$ represent the amount of the other tea. But you could also let x represent the amount of one tea and y represent the amount of the other, and include the equation $x + y = 10$ as one equation in your system.

In theory, you could do that with any problem with two unknowns with one variable if you know how the unknowns are related, but that's easy when you know they add to 10. It's not so easy when you know that 17 of these and 32 of those cost $107.32. There are times when it's just easier to use two variables.

Suppose a group of friends go out to get some fast food, and split into two groups in two different lines while waiting to order. The first group bought 8 slices of pizza and 6 soft drinks for $41.14. The second group bought 5 slices of pizza and 6 soft drinks for $30.10. How much does one slice of pizza cost?

$$\text{Let } x = \text{the price of a slice of pizza}$$

$$\text{Let } y = \text{the price of a soft drink}$$

$$8 \text{ slices of pizza and 6 soft drinks cost } \$41.14$$

$$8x + 6y = \$41.14$$

$$5 \text{ slices of pizza and 6 soft drinks for } \$30.10$$

$$5x + 6y = \$30.10$$

$$\text{Solve the system } \begin{cases} 8x + 6y = \$41.14 \\ 5x + 6y = \$30.10 \end{cases}$$

$$8x + 6y = \$41.14$$

$$\underline{5x + 6y = \$30.10}$$

$$3x = 11.04$$

$$x = 3.68$$

$$5(3.68) + 6y = 30.10$$

$$18.40 + 6y = 30.10$$

$$6y = 11.70$$

$$y = 1.95$$

A slice of pizza costs $3.68 and a soft drink costs $1.95.

EXERCISE 14·5

1. A collection of dimes and nickels is made up of 16 coins and is worth $1.25. How many nickels are in the collection?

2. Adam made a trip to a town 138 miles away. He needed to walk to a bus stop and then ride the bus to the town. The total trip took 6 hours. If Adam walks at 4 miles per hour and the bus has an average speed of 28 mph, how long did Adam walk?

3. Mrs. Connolly invested a total of $6,000 in two investments. One portion was invested at 4% and the remainder at 7%. If her total earnings were $342, find the amount invested at each rate.

4. Chrissy and her younger sister, Claire, play a game to help Claire learn her math facts. Chrissy gives Claire 30 math problems. For each problem Claire gets correct, Chrissy gives her 7 cents, but each time Claire is wrong, she has to give Chrissy back 4 cents. If Claire ends the game with 12 cents, how many problems did she get correct?

5. If the first of two numbers is increased by 10% of the second, the result is 17. If the second number is increased by 25% of the first, the result is 14. Find the numbers.

6. How many pounds of a tea that sells for $1.60 per pound and a tea that sells for $3.20 per pound should be mixed to create 40 pounds of a blend that can be sold for $2.00 per pound?

7. Laura Anne works as a web designer and as an editor. She earns $250 for each web page she creates, and $50 for each page she edits. Last week she worked on a total of 28 pages between these two jobs. If she earned $3,200 last week, how many web pages did Laura Anne produce?

8. A farmer kept pigs and chickens. He has 234 animals, and the animals have a total of 560 legs. How many chickens does he have?

9. Whenever there is a bake sale, Pat makes brownies and cupcakes, 48 pieces total, and they always sell out. Brownies are sold for $1.50 each, and cupcakes for $2.25. At the January sale, Pat's baked goods sold for a total of $81.00. How many brownies did Pat make?

10. A collection of nickels and dimes is worth $1.35. If the number of nickels and the number of dimes are interchanged, the value would be $1.65. What is the total number of coins?

Answers

1   Arithmetic to algebra

1·1
1. Rationals, Reals
2. Integers, Rationals, Reals
3. Rationals, Reals
4. Whole, Integers, Rationals, Reals
5. Irrationals, Reals
6. Natural, Whole, Integers, Rationals, Reals
7. Rationals, Reals
8. Rationals, Reals
9. Rationals, Reals
10. Irrationals, Reals

11–20.

```
        G   I  F  C     D     B A   E H J
   <----•---•--•--•--+--•--+--+-•-+-•-•-•---->
      -10-9-8-7-6-5-4-3-2-1 0 1 2 3 4 5 6 7 8 9 10
```

1·2
1. Commutative Property for Addition
2. Associative Property for Multiplication
3. Identity for Addition
4. Inverse for Multiplication
5. Distributive Property
6. Zero Product Property
7. Associative Property for Addition
8. Identity for Multiplication
9. Commutative Property for Multiplication
10. Multiplication Property of Zero
11. Inverse for Addition
12. Distributive Property
13. Identity for Multiplication
14. Associative Property for Addition
15. Inverse for Addition

1·3
1. 2
2. −17
3. −54
4. 4
5. 3
6. 14
7. 24
8. −3
9. −8
10. 18
11. 2
12. 7
13. −32
14. −10
15. −40
16. −12
17. −1
18. −48
19. 16
20. 5

1·4
1. 9
2. 225
3. 10
4. 4
5. 23
6. 21
7. 3
8. 15
9. 20
10. −2

1·5
1. $11t$
2. $4x$
3. $3x+3y$
4. $x+10y-3$
5. $-1+2x-2x^2$
6. $13t-3r-10$
7. $7x^2-6x+19$
8. $8x-6y-19$
9. $2x^2+2x+1$
10. $10y-9x$
11. $2+3x$
12. $3y-7$
13. $\dfrac{t}{3}+11$
14. $9n-8$
15. $w+(-w)$
16. $\dfrac{5p-3}{p^2}$
17. r^2-4r
18. $\dfrac{x}{2x-1}+8$
19. $(3z+2)(4z-6)$
20. $\sqrt{4v^2-1}$

1·6
1. 14
2. −16
3. 1
4. 2
5. 15
6. 1
7. 85
8. 230
9. −3
10. −26

2 Linear equations

2·1
1. $x = 4$
2. $y = 16$
3. $t = 3$
4. $w = 37$
5. $x = 2$
6. $z = 13.1$
7. $y = 12\frac{1}{10}$
8. $x = -6$
9. $y = 3$
10. $t = -7$

2·2
1. $x = 4$
2. $z = 63$
3. $y = -8\frac{2}{5}$
4. $t = -36$
5. $x = 30$
6. $w = 15.4$
7. $t = \dfrac{75}{128}$
8. $m = -12.4$
9. $x = -3$
10. $z = 175$

2·3
1. $x = 13$
2. $t = -3$
3. $x = 5$
4. $x = \dfrac{1}{3}$
5. $x = 12$
6. $x = 7$
7. $x = 16$
8. $x = 0.5$
9. $x = 25$
10. $x = -8$

2·4
1. $x = 5$
2. $x = -4$
3. $17 = x$
4. $-1 = x$
5. $x = \dfrac{4}{3}$
6. $x = 5.8$
7. $x = 2.5$
8. $x = 31$
9. $\dfrac{13}{8} = x$
10. $x = -\dfrac{3}{2}$

2·5
1. $x = 6$
2. $x = 10$
3. $11 = x$
4. $x = 0$
5. $11 = x$
6. $2 = x$
7. $x = 6$
8. $x = \dfrac{8}{3}$
9. $x = \dfrac{1}{3}$
10. $x = 3$

2·6
1. $x = 6 \quad x = \dfrac{-28}{3}$
2. $x = \dfrac{10}{3} \quad x = -\dfrac{7}{3}$
3. $x = 9 \quad x = -9.8$
4. $x = 5 \quad x = -6$
5. $x = 5.5 \quad x = -7.25$
6. $x = 2$ Reject $x = -\dfrac{10}{7}$ because it will make the 18x negative,
7. $x = 5$ (Reject $x = -10$)
8. $x = 9.5 \quad x = 0.75$
9. $x = 4 \quad x = -2$
10. $x = 3 \quad x = -\dfrac{19}{13}$

2·7
1. 5 nickels
2. $4\frac{1}{6}$ h, or 4 h and 10 min, later
3. $3\frac{1}{3}$ lb of peanuts and $6\frac{2}{3}$ lbs of raisins
4. 23 dimes
5. 5:00 p.m.
6. 250 pennies
7. 481 students
8. 12 ounces of Sweet Rose Tulsi tea and 4 ounces of Orange Blossom green tea
9. 1:15 p.m.
10. 2:30 p.m.

3 Linear inequalities

3·1
1. $x \geq 9$

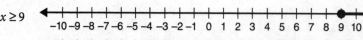

2. $x > 3$

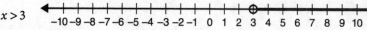

3. $x \ge -4$

4. $x < 3$

5. $t \ge 3$

6. $y > \dfrac{7}{2}$

7. $x \ge 7$

8. $x \ge -4$

9. $x < -1$

10. $x \le \dfrac{7}{2}$

3·2

1. $x > 40$ or $x < -6$

2. $15 < y < 24$

3. $-2 \le x < -\dfrac{3}{4}$

4. $x \ge 5$ or $x < 2$

5. $-4 < x \le 1$

6. $x > -\dfrac{2}{3}$ or $x < -9$

7. $-1 \le y \le \dfrac{1}{15}$

8. $20 > x > \dfrac{49}{3}$

9. $x > -1$ or $x > 10$, which is equivalent to $x > -1$

10. $y > 3$ and $y \ge 7$, which is equivalent to $y \ge 7$

3·3

1. $-1 < x < 8$

2. $-\dfrac{22}{3} \ge x$ or $x \ge 4$

3. $-\dfrac{37}{3} > x$ or $x > 9$

4. $1.8 \geq x \geq -0.2$

(number line from -2 to 2, closed circles at 0 and ~1.8)

5. $\frac{3}{7} > x$ or $x > 3$

(number line from -5 to 5, open circles at ~0.4 and 3)

6. $15 < x$ or $x < 3$

(number line from -2 to 18, open circles at 3 and 15)

7. $-\frac{43}{11} \geq x$ or $x \geq 3$

(number line from -10 to 10, closed circles at ~-4 and 3)

8. $12 \leq x$ or $x \leq 5$

(number line from -20 to 20, closed circles at ~6 and 12)

9. $1.6 < x$ and $x < 2$

(number line from -2 to 2, open circles at ~1.6 and 2)

10. $x \geq 4$ or $x \leq 2$

(number line from -10 to 10, closed circles at 2 and 4)

4   Coordinate graphing

4·1   1–5.

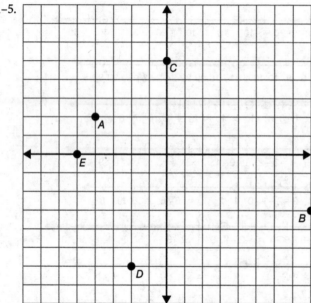

6. Quadrant II   9. Quadrant III
7. Quadrant I   10. Quadrant IV
8. Quadrant IV

4·2
1. $d = 3\sqrt{10}$ 6. $a = 4,\quad a = 10$
2. $d = \sqrt{17}$ 7. $d = 15,\quad d = -9$
3. $d = \sqrt{29}$ 8. $c = 15,\quad c = 1$
4. $d = \sqrt{34}$ 9. $b = -10,\quad b = 8$
5. $d = 7$ 10. $a = \pm 4$

4·3
1. $(3.5, 5.5)$ 6. $x = 2$
2. $(-2, 4.5)$ 7. $x = 7$
3. $(-3, -2)$ 8. $y = 9$
4. $(4, 4)$ 9. $x = -5$
5. $(2, -3)$ 10. $x = 16$

4·4

1. $m = -\dfrac{3}{5}$ 6. $y = -2$

2. $m = -\dfrac{2}{3}$ 7. $x = 4$

3. $m = 0$ 8. $y = 4.5$

4. $m = \dfrac{1}{4}$ 9. $x = -8$

5. Undefined   10. $y = 3$

4·5

1.

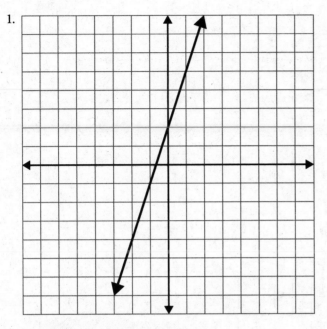

x	−2	−1	0	1	2
y	−4	−1	2	5	8

2.

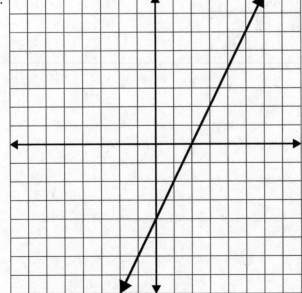

x	−2	−1	0	1	2
y	−8	−6	−4	−2	0

3.

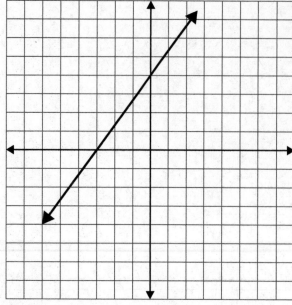

x	−6	−3	0	3	6
y	−4	0	4	8	12

4.

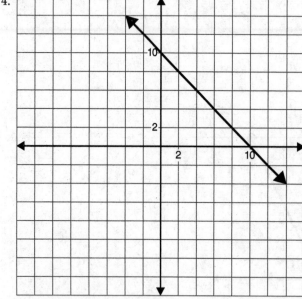

x	−2	−1	0	1	2
y	12	11	10	9	8

5.

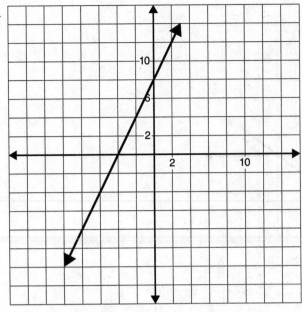

x	−2	−1	0	1	2
y	3	5	7	9	11

6.

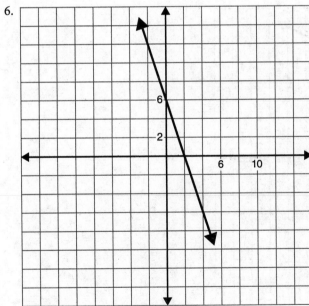

x	−2	−1	0	1	2
y	12	9	6	3	0

7.

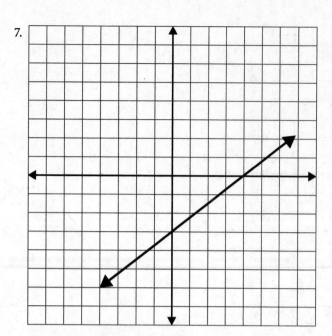

x	−8	−4	0	4	8
y	−9	−6	−3	0	3

8.

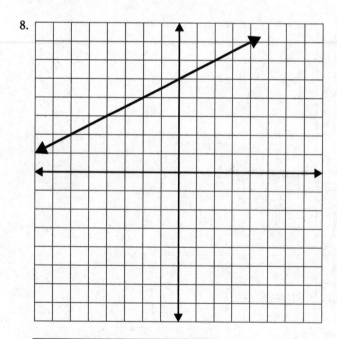

x	−4	−2	0	2	4
y	3	4	5	6	7

9.

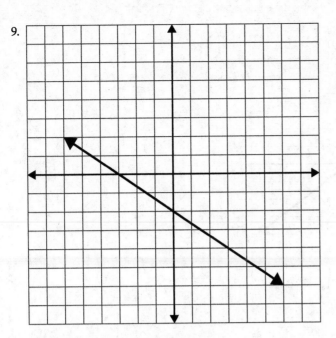

x	−6	−3	0	3	6
y	2	0	−2	−4	−6

10.

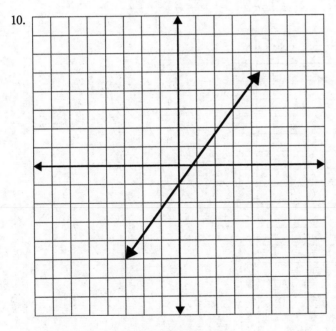

x	−6	−3	0	3	6
y	−9	−5	−1	3	7

4·6

1.

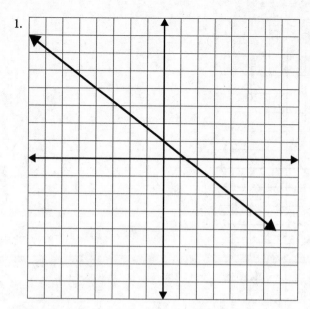

2.

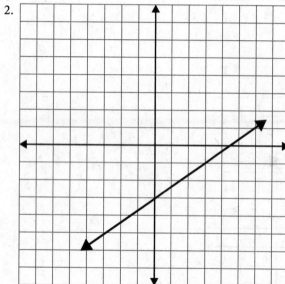

3.

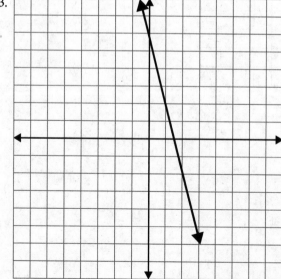

4.

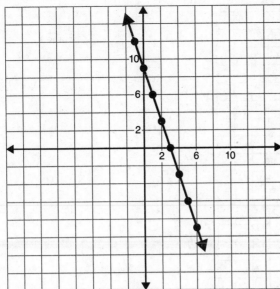

5.

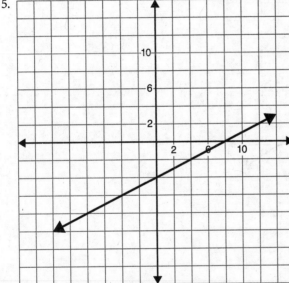

6.

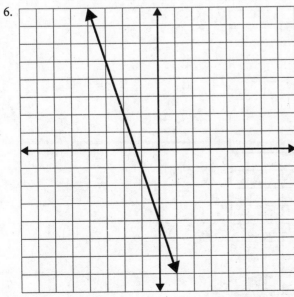

7.

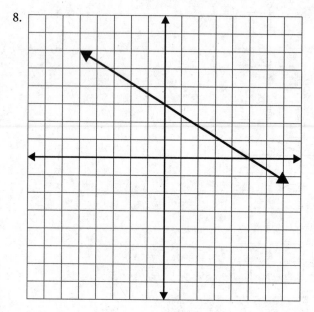

8.

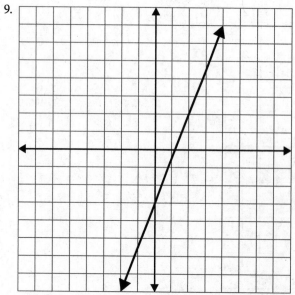

9.

10.

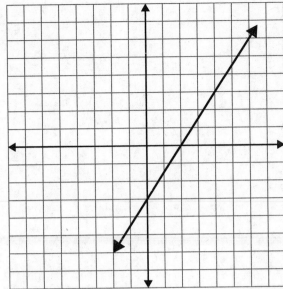

4·7   1. Vertical
2. Horizontal
3. Vertical
4. Oblique
5. Horizontal

6.

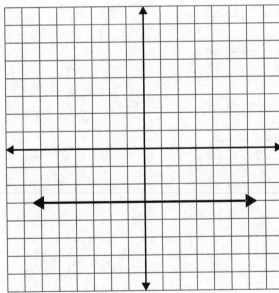

7.

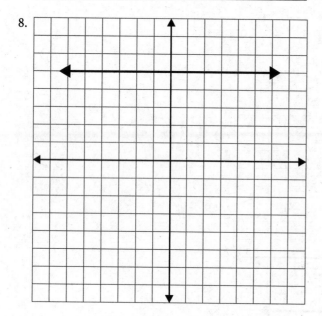

8.

9.

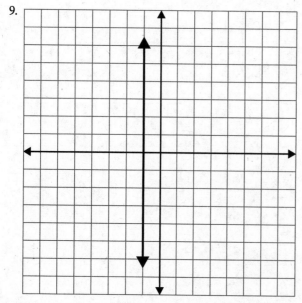

10.

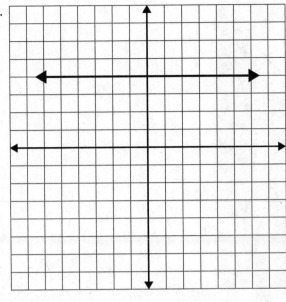

4·8   1.

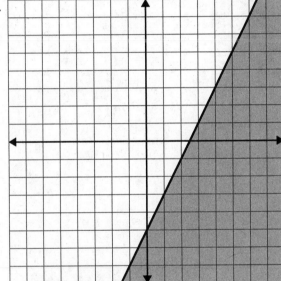

2.

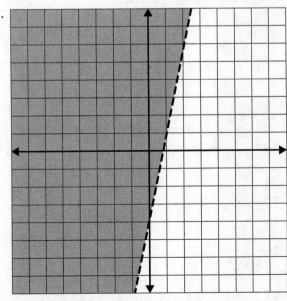

3.

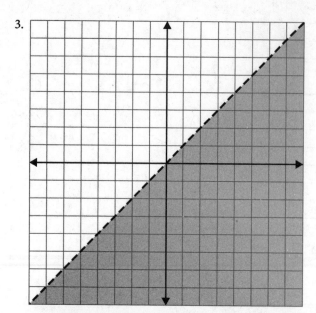

4.

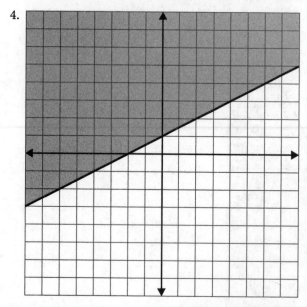

5.

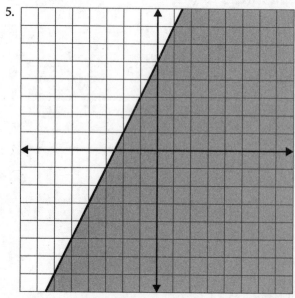

6.

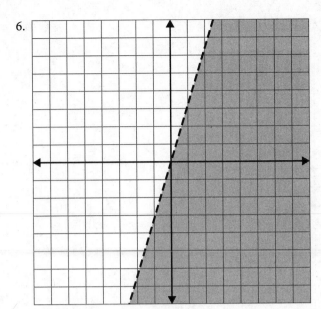

7.

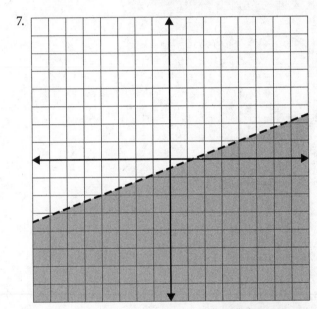

8.

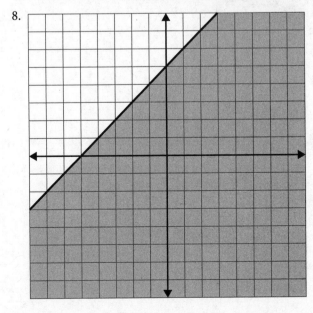

9.

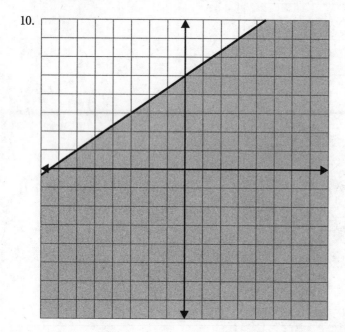

10.

1.

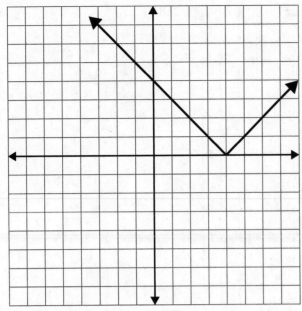

x	2	3	4	5	6
y	2	1	0	1	2

2.

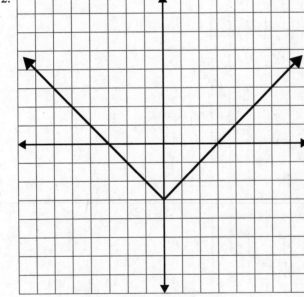

x	−2	−1	0	1	2
y	−1	−2	−3	−2	−1

3.

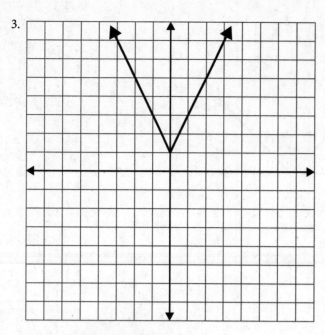

x	−2	−1	0	1	2
y	5	3	1	3	5

4.

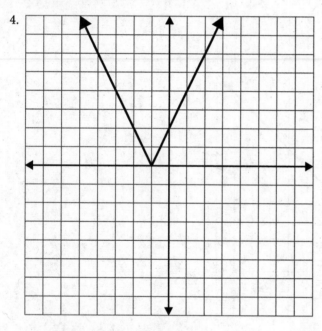

x	−3	−2	−1	0	1
y	4	2	0	2	4

5.

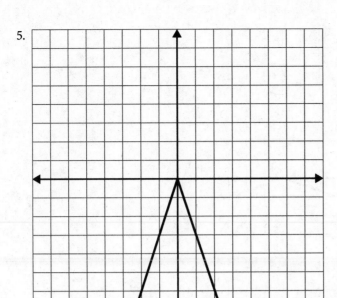

x	−2	−1	0	1	2
y	−6	−3	0	−3	−6

6.

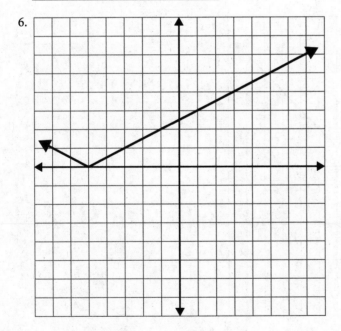

x	−9	−7	−5	−3	−1
y	2	1	0	1	2

7.

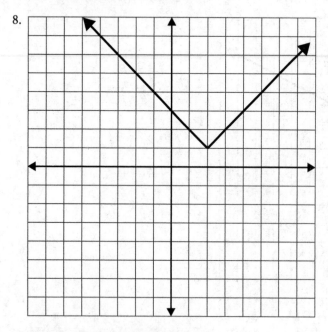

x	−1	2	5	8	11
y	−2	−1	0	−1	−2

8.

x	0	1	2	3	4
y	3	2	1	2	3

9.

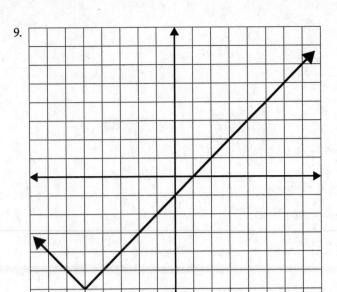

x	−7	−6	−5	−4	−3
y	−4	−5	−6	−5	−4

10.

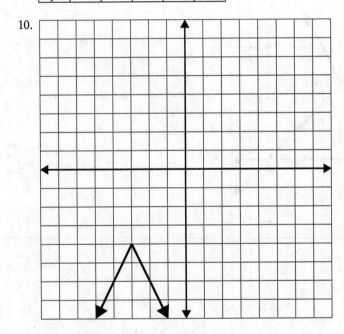

x	−5	−4	−3	−2	−1
y	−8	−6	−4	−6	−8

4·10

1. $y = 3x + 8$

2. $y = -5x + 2$

3. $y = \dfrac{2}{3}x + 6$

4. $y = 4x - 5$

5. $y = \dfrac{1}{2}x + 1$

6. $y = -\dfrac{3}{2}x - 7$

7. $y = 2x + 3$

8. $y = \dfrac{2}{3}x + 2$

9. $y = \dfrac{4}{3}x - 3$

10. $y = -\dfrac{3}{2}x + 6$

4·11

1. Perpendicular

2. Parallel

3. Neither

4. Parallel

5. Perpendicular

6. $y = 5x - 16$

7. $y = -\dfrac{3}{4}x + 2$

8. $y = -\dfrac{4}{3}x + \dfrac{7}{3}$

9. $y = -\dfrac{1}{4}x + 14$

10. $y = 2x - 16$

5 Systems of linear equations and inequalities

5·1

1.

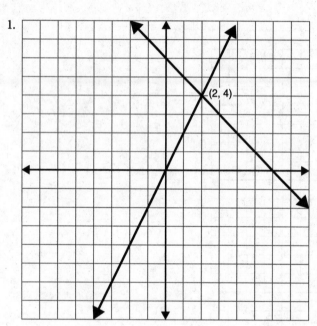

(2, 4)

2.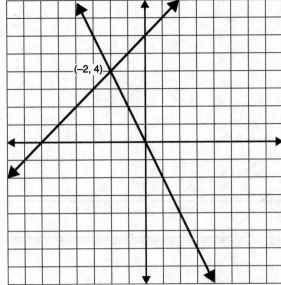

(−2, 4)

3.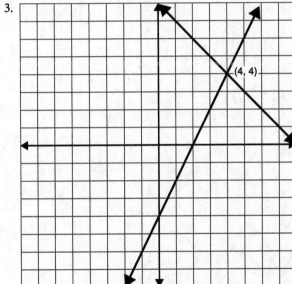

(4, 4)

4.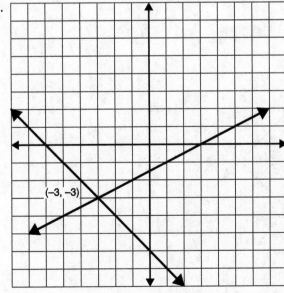

(−3, −3)

5.

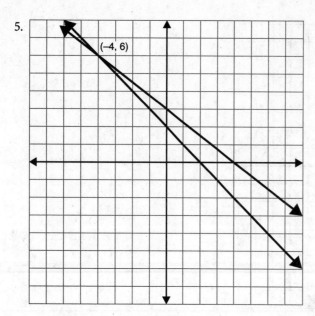

(−4, 6)

6.

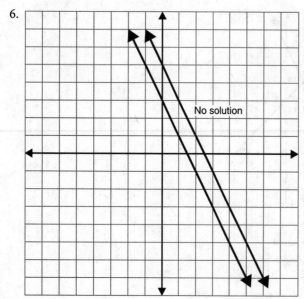

No solution

7.

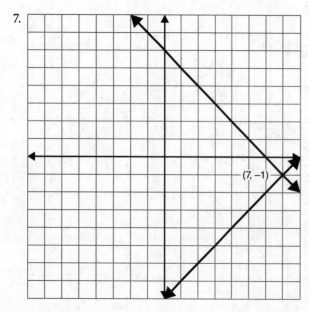

(7, −1)

8.

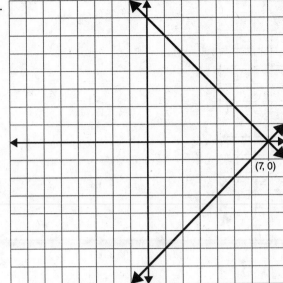

(7, 0)

9.

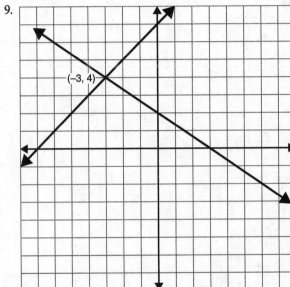

(−3, 4)

10.

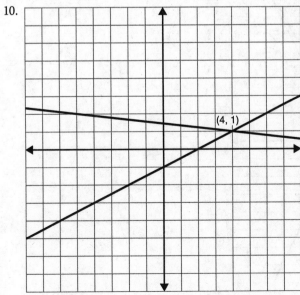

(4, 1)

5·2

1.

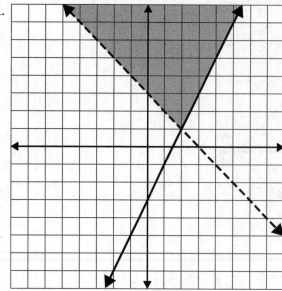

2.

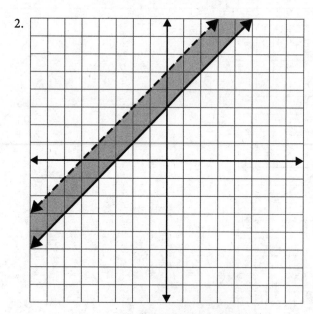

3.

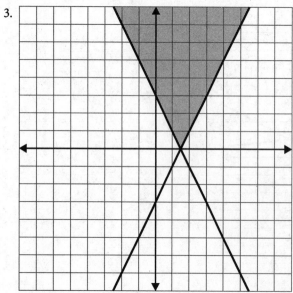

4.

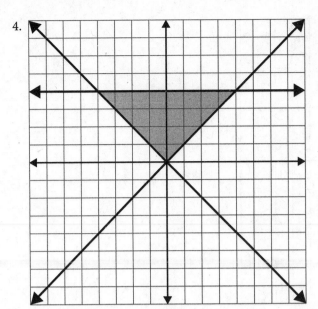

5.

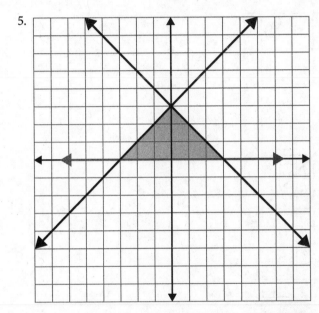

6.

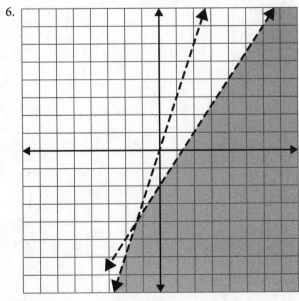

7.

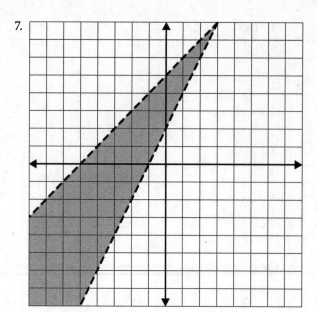

8.

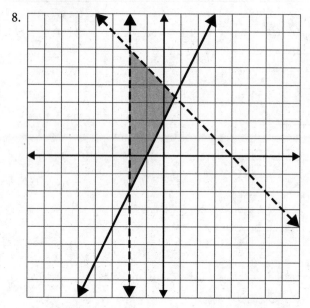

9.

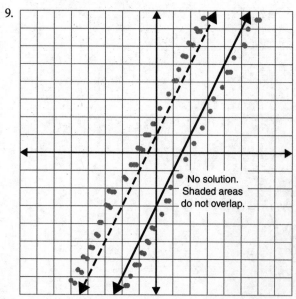

No solution.
Shaded areas
do not overlap.

10.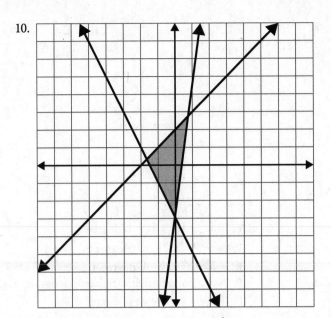

5·3

1. $x = 5, y = 5$
2. $x = 4, y = 8$
3. $x = 3, y = 9$
4. $x = 30, y = 27$
5. $x = 19, y = 23$

6. $x = 54, y = 8$
7. $x = 3, y = 1$
8. $x = 9, y = 4$
9. $x = 11, y = 2$
10. $x = -1, y = 5$

5·4

1. $x = 6, y = 2$
2. $x = 10, y = 7$
3. $x = 4, y = 1$
4. $x = 7, y = 1$
5. $x = 2, y = \dfrac{1}{2}$

6. $x = 2, y = 3$
7. $x = 3, y = 0$
8. $x = 3, y = 10$
9. $x = 8, y = -7$
10. $x = 3, y = \dfrac{3}{2}$

5·5

1. $a = 1, b = 6$
2. $x = 1, y = 10$
3. $x = 2.5, y = -2$
4. $x = 7, y = 3$
5. $x = 5, y = 2$

6. $x = \dfrac{1}{5}, y = \dfrac{6}{5}$
7. $x = 3, y = -3$
8. $x = -2, y = -5$
9. $x = -\dfrac{1}{6}, y = \dfrac{4}{3}$
10. $x = 0.6, y = 6.6$

5·6

1. Dependent
2. Inconsistent
3. Consistent
4. Inconsistent
5. Dependent

6. Consistent
7. Consistent
8. Inconsistent
9. Consistent
10. Dependent

6 Powers and polynomials

6·1

1. x^{11}
2. y^6
3. $6x^6$
4. $21x^{10}$
5. x^6

6. $t^{-1} = \dfrac{1}{t}$
7. y^7
8. $x^{-5} = \dfrac{1}{x^5}$
9. x^2
10. y^{21}

11. $x^0 = 1$
12. x^{10}
13. x^2
14. $t^{-1} = \dfrac{1}{t}$
15. x^6

6·2

1. $4x^{10}$

2. $-8x^9$

3. $20a^8$

4. $-27x^5y^{15}$

5. $72b^{11}$

6. $\dfrac{x^{10}}{9}$

7. $\dfrac{4}{25}x^4$

8. $\dfrac{256}{81t^8}$

9. $\dfrac{-64x^3y^{12}}{9}$

10. $\dfrac{81x^8}{16y^4}$

6·3

1. $2x^3+3x^2+5x-7$; degree 3
2. $5t^{12}+t^7+8t^2-9t-1$; degree 12
3. $-12y^{11}+5y^6-2y^3+8$; degree 11
4. Not a polynomial; variable under radical
5. $2x^5-4x^3+3x$; degree 5
6. $-3z^7-4z^2+8z+4$; degree 7
7. w^5-9w^3-3w+7; degree 5
8. $-b^4+b^2-3b-4$; degree 4
9. Not a polynomial; variable in denominator
10. $-7y^3+8y^2-4y+6$; degree 3

6·4

1. $14w^2-9w-1$

2. $2a^2-5a-4$

3. $-9x^2+41x-24$

4. $-4y^2-3y+32$

5. $4-b+4b^2$

6. $4b^2-3b+3$

7. $11x^2-13x+2$

8. $-2x^2-7x+2$

9. $-3x^2+x+2$

10. $2x^2-16x+3$

6·5

1. $-6b^7$

2. $30x^4y^4$

3. $-36x^5y^2z^{10}$

4. $-3a^2b^2c^3$

5. $40a^3b$

6. $18x^6y^3$

7. $-36w^5x^6$

8. $4x^6$

9. $20b^8$

10. $-27r^3t^9$

11. $(2x^3)(\underline{-3x^2})=-6x^5$

12. $(-3b^2)(\underline{-4b^5})=12b^7$

13. $(\underline{-5x^4y^2})(3x^2y)=-15x^6y^3$

14. $(\underline{-3x^2z})(-2z^4)=6x^2z^5$

15. $(6xy^2)\left(-\dfrac{1}{\underline{2}}x^2y\right)=-3x^3y^3$

6·6

1. $10a^3+15a^2$

2. $-2x^4+6x^3+4x^2$

3. $22y^4-6y^3+10y^2$

4. $-6b^5+9b^4-12b^3$

5. $3x^3y+5x^2y^2-2xy^3$

6. $25x^4y-35x^3y^2+5x^2y^3$

7. $8x^2+16xy-24xz$

8. $-5a^3b+5ab^4$

9. $4x^{10}-3x^8+5x^7-x^5+7x^4-10x^3$

10. $9a^6b^4c^2-6a^4b^4c^3+21a^9b^3c^6$

11. $\underline{3}(x+1)=3x+3$

12. $a(\underline{b-5})=ab-5a$

13. $\underline{4}(2x-y)=8x-4y$

14. $7x(\underline{1+7x})=7x+49x^2$

15. $\underline{2ab}(2a+b)=4a^2b+2ab^2$

6·7

1. $x^2+10x+16$

2. $y^2-13y+36$

3. $t^2+4t-12$

4. $2x^2+2x-24$

5. $3y^2-26y-9$

6. $15x^2+2x-24$

7. $6x^2+29x-5$

8. $5-13b-6b^2$

9. $6x^2+x-35$

10. $-10x^2+29x-10$

11. x^2-16

12. x^2-9

13. $4x^2-1$

14. $9x^2-25$

15. $49-9x^2$

16. $(x+3)(x+\underline{2})=x^2+5x+6$

17. $(x-7)(x-\underline{2})=x^2-9x+14$

18. $(2a+1)(a+\underline{4})=2a^2+9a+4$

19. $(3x-2)(x-\underline{5})=3x^2-17x+10$

20. $(2t+3)(3t-\underline{5})=6t^2-t-15$

6·8

1. $2a^3 + 19a^2 + 38a + 21$

2. $6b^3 + 13b^2 + 16b + 15$

3. $4c^3 - 39c^2 + 54c + 16$

4. $8x^4 - 18x^3 + 17x^2 - 19x + 7$

5. $y^3 - y^2 - 19y + 4$

6. $x^3 - 6x^2 + 12x - 8$

7. $t^3 - 8$

8. $x^3 + 1$

9. $x^4 + 4x^3 - 16x - 16$

10. $2y^4 - y^3 - 8y^2 + 34y - 15$

6·9

1. $8c^3d^2$

2. $5d^4$

3. $4x^7$

4. $5y^4$

5. $-4x^2y$

6. 1

7. $-3t^2$

8. $-18r^2$

9. $8x - 9x^4$

10. $5y^3 - 7y^5$

11. $x + 3$

12. $9 - 13y$

13. $8z^8 - 7z^7 + 6z^6 - 5z^5$

14. $3x^3y^2 - 9x^2y^3 + 27xy^4$

15. $3x^2 - x + 4$

6·10

1. $x - 7$

2. $y + 4$

3. $2x + 3$

4. $7x^2 + 5$

5. $3x - 6 + \dfrac{-3}{3x - 8}$

6. $2a + 1 + \dfrac{2}{a + 3}$

7. $2b - 1$

8. $4x^2 - x - 5 + \dfrac{5}{3x + 5}$

9. $x^2 + 6$

10. $4y^2 - 2y + 1$

7 Factoring

7.1

1. $2x^2$

2. $5r$

3. 11

4. $2x$

5. ax

6. $8t^3$

7. az

8. $3y^2$

9. $2a$

10. 3

7·2

1. $y(y - 15)$

2. $3b(b - 2)$

3. $8ab(4a + 5)$

4. $5(y^2 + 3y + 4)$

5. $x^3y^4(x^5 - xy^3 + y)$

6. $-a^2(1 + a - 2a^2)$

7. $25x^4(1 - 2x + 5x^3)$

8. $8r(r + 3t + 2)$

9. $16x^2y(y - 3x)$

10. $3xy(x + 2 + 5xy)$

7·3

1. $(x + 5)(x + 7)$

2. $(x + 4)(x + 7)$

3. $(x - 3)(x - 5)$

4. $(x - 3)(x - 4)$

5. $(x + 5)(x - 4)$

6. $(x - 3)(x + 1)$

7. $(x - 9)(x - 2)$

8. $(x - 11)(x + 2)$

9. $(x + 13)(x - 3)$

10. $(x + 4)(x + 8)$

7·4

1. $(3x + 5)(x + 2)$

2. $(2x - 1)(x - 1)$

3. $(2x + 1)(x + 3)$

4. $(6x + 1)(2x + 5)$

5. $(2x + 3)(3x + 4)$

6. $(10x + 1)(x - 5)$

7. $(3x - 5)(3x - 4)$

8. $(3x + 2)(6x + 1)$

9. $(3x + 1)(5x - 6)$

10. $(x - 6)(4x - 5)$

7·5

1. $(x + 7)(x - 7)$

2. $(x + 3)^2$

3. $(6t + 1)(6t - 1)$

4. $(3t - 4)^2$

5. $(4 + y)(4 - y)$

6. $(3y + 7)^2$

7. $(2x + 9)(2x - 9)$

8. $(2x + 1)^2$

9. $(4a + 3y)(4a - 3y)$

10. $(x - 6)^2$

7·6   1. $(x+1)(x-1)(x+2)$ 6. $(z+2)(2x+3)$

   2. $(x+2)(x-2)(x-3)$ 7. $(y+1)(5x+3)$

   3. $(x+10)(x-10)(x+2)$ 8. $(y^2-4)(x^2-1)=(x+1)(x-1)(y+2)(y-2)$

   4. $(x^2+4)(2x-1)$ 9. $(x^2+1)(y^2-9)=(x^2+1)(y+3)(y-3)$

   5. $(y-3)(3x+5)$ 10. $(y^2+4)(x^2+1)$

8   Radicals

8·1   1. 6   6. 12

   2. −9   7. −2

   3. ±5   8. −2

   4. 3   9. 10

   5. −5   10. No Real Root

8·2   1. $3<\sqrt{11}<4$, $\sqrt{11}\approx 3.3$ 6. $4<\sqrt{20}<5$, $\sqrt{20}\approx 4.5$

   2. $9<\sqrt{90}<10$, $\sqrt{90}\approx 9.5$ 7. $2<\sqrt[3]{10}<3$, $\sqrt[3]{10}\approx 2.2$

   3. $12<\sqrt{150}<13$, $\sqrt{150}\approx 12.2$ 8. $2<\sqrt[3]{16}<3$, $\sqrt[3]{16}\approx 2.5$

   4. $6<\sqrt{48}<7$, $\sqrt{48}\approx 6.9$ 9. $3<\sqrt[3]{30}<4$, $\sqrt[3]{30}\approx 3.1$

   5. $1<\sqrt{3}<2$, $\sqrt{3}\approx 1.7$ 10. $-5<\sqrt[3]{-81}<-4$, $\sqrt[3]{-81}\approx -4.3$

8·3   1. $4\sqrt{2}$ 6. $2y\sqrt{3y}$

   2. $6\sqrt{2}$ 7. $5b^2\sqrt{2ab}$

   3. $a\sqrt{a}$ 8. $3xy\sqrt{3x}$

   4. $7\sqrt{2}$ 9. $7a\sqrt{a}$

   5. $2x\sqrt{2}$ 10. $4a^3bc\sqrt{3ab}$

8·4   1. $\dfrac{\sqrt{6}}{2}$ 6. $\dfrac{-9}{2}\left(\sqrt{5}+3\right)$

   2. $\dfrac{24\sqrt{5}}{5}$ 7. $-5\left(1+\sqrt{5}\right)$

   3. $\sqrt{5}$ 8. $\dfrac{-16+5\sqrt{7}}{9}$

   4. $\sqrt{3}$ 9. $\dfrac{3}{2}\left(\sqrt{5}+\sqrt{3}\right)$

   5. 10   10. $-11-6\sqrt{3}$

8·5   1. $6\sqrt{3}$ 6. $4\sqrt{7}$

   2. $3\sqrt{5}$ 7. $-3\sqrt{11}$

   3. $\sqrt{12}$ 8. $10\sqrt{6}$

   4. $12\sqrt{2}$ 9. $4\sqrt{3}$

   5. $7\sqrt{5}$ 10. $10\sqrt{5}$

8·6
1. $x = 9$
2. No solution
3. $x = 81$
4. No solution
5. $y = 50$

6. No solution
7. $x = 17$
8. $x = 26$
9. $x = 11$
10. $x = \dfrac{13}{9}$

8·7
1. $x = 5$
2. $x = 5$
3. $x = 4$
4. $x = -1.25$
5. $x = 3$

6. $x = \dfrac{109}{8}$
7. $x = 9$
8. $x = 25$
9. $x = -1.75$
10. $x = 11$

8·8

1.
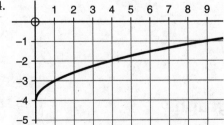

2.
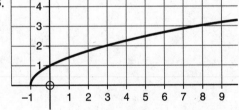

3.
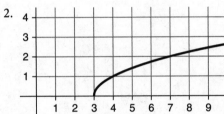

4.
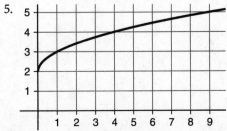

5.

6.

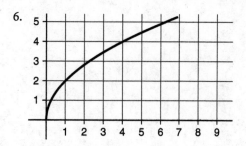

7.

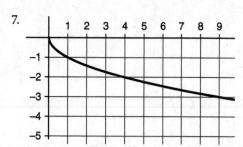

8.

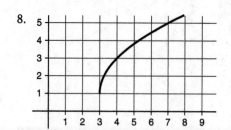

9.

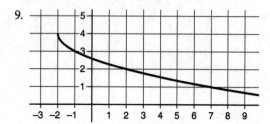

10.

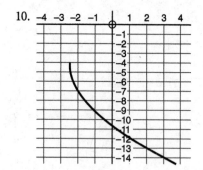

9   Quadratic equations and their graphs

9·1

1. $x = \pm 8$

2. $x = \pm 4$

3. $x = \pm 5$

4. $x = \pm 3\sqrt{2}$

5. $x = \pm 4$

6. $t = \pm 10\sqrt{10}$

7. $y = \pm 5\sqrt{3}$

8. $x = \pm \dfrac{2}{3}$

9. $y = \pm \dfrac{5}{8}$

10. $x = \pm 3\sqrt{3}$

9·2

1. $x = 7, x = -3$
2. $x = 2, x = -4$
3. $x = 3 \pm 4\sqrt{3}$
4. $x = -1 \pm 5\sqrt{3}$
5. $x = \dfrac{5 \pm 2\sqrt{3}}{3}$

6. $y = 4 \pm \sqrt{23}$
7. $x = 7, x = 2$
8. $x = -2 \pm 2\sqrt{2}$
9. $a = \dfrac{-5 \pm \sqrt{37}}{2}$
10. $t = 5 \pm \sqrt{17}$

9·3

1. $x = 3, x = -7$
2. $t = 2, t = -5$
3. $y = 8, y = -4$
4. $x = 3, x = -2$
5. $x = -3 \pm 3\sqrt{2}$

6. $t = -3 \pm 2\sqrt{6}$
7. $x = 1, x = -\dfrac{3}{4}$
8. $x = 1, x = -\dfrac{1}{3}$
9. $x = -1, x = \dfrac{5}{3}$
10. $x = \dfrac{2}{3}, x = -\dfrac{1}{2}$

11. $b^2 - 4ac = 61$, two irrational solutions
12. $b^2 - 4ac = -11$, no real solutions
13. $b^2 - 4ac = 25$, two rational solutions
14. $b^2 - 4ac = 0$, one rational solution
15. $b^2 - 4ac = 24$, two irrational solutions

16. $b^2 - 4ac = 1$, two rational solutions
17. $b^2 - 4ac = 44$, two irrational solutions
18. $b^2 - 4ac = -31$, no real solutions
19. $b^2 - 4ac = 0$, one rational solution
20. $b^2 - 4ac = 69$, two irrational solutions

9·4

1. $x = -2, x = -3$
2. $x = 4, x = 3$
3. $y = 2, y = -4$
4. $a = 5, a = -2$
5. $x = 4, x = -5$

6. $x = 5, x = 1$
7. $x = 0, x = -3$
8. $x = 0, x = 5$
9. $x = 1, x = \dfrac{1}{2}$
10. $x = 3, x = -\dfrac{5}{2}$

9·5

1. $(3, 0), (1, 0), (0, 3)$
2. $(5, 0), (-1, 0), (0, -5)$
3. $(-2, 0), (0, 0)$
4. $(3, 0), (4, 0), (0, 12)$
5. $(-1, 0), \left(\dfrac{1}{2}, 0\right), (0, 1)$

6. $x = 4, (4, -1)$
7. $x = -2, (-2, -6)$
8. $x = 1, (1, 1)$
9. $x = 3, (3, 2)$
10. $x = 2, (2, 11)$

9·6

1.

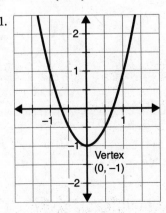

Vertex
$(0, -1)$

2.

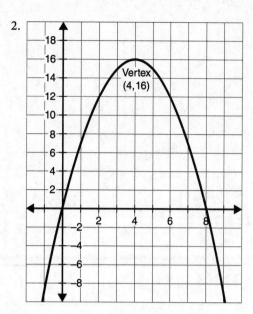

Vertex
(4, 16)

3.

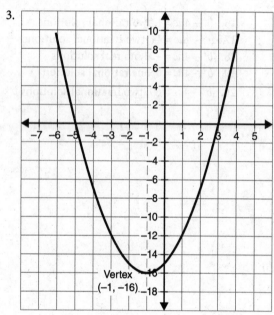

Vertex
(−1, −16)

4.

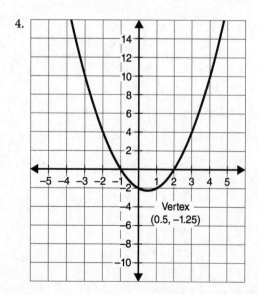

Vertex
(0.5, −1.25)

5.

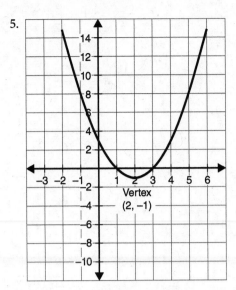

Vertex
(2, −1)

6.

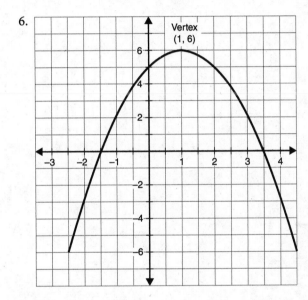

Vertex
(1, 6)

7.

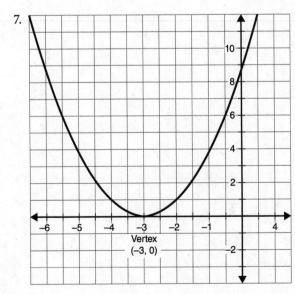

Vertex
(−3, 0)

8.

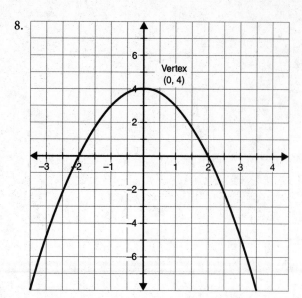

Vertex
(0, 4)

9.

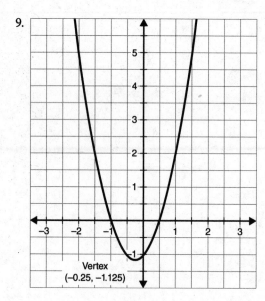

Vertex
(−0.25, −1.125)

10.

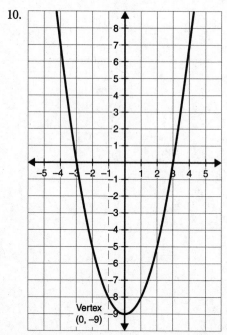

Vertex
(0, −9)

10 Proportion and variation

10·1
1. 45°, 60°, 75°
2. 8 ft. and 12 ft.
3. 104 and 39
4. 20 oz
5. 24, 32, and 64
6. 20 and 24
7. 6 and 14
8. 51
9. 8 and 20
10. 6 and 20

10·2
1. $x = 4.2$
2. $w = 15$
3. $x = 25$
4. $x = 4\frac{1}{6}$
5. $x = 15$
6. $x = 29$
7. $x = \pm 8$
8. $x = \pm 5$
9. $x = \pm 13$
10. $x = 7, x = -4$

10·3
1. 42
2. 100
3. 312
4. 1275
5. 700
6. 0.009 V
7. 143 mi
8. 0.6 mi
9. 304 cm^3
10. 70 mi

10·4
1. $k = 3, y = 3$
2. $k = 32, x = 16$
3. $k = 72, y = 8$
4. $k = 132, t = 44$
5. $k = 540, a = 12$
6. $k = 128, v = \dfrac{16}{3} = 5\frac{1}{3}$
7. $k = 88, w = 4$
8. $k = 24, B = \dfrac{1}{6}$
9. $k = 234, x = 13$
10. $k = 288, a = 24$

10·5
1. $y = 180$
2. $x = 8$
3. $z = 7$
4. $y = 5$
5. $x = 21$
6. $z = 42$
7. 20 cm
8. Approximately 683 N
9. Approximately 7.8 ft^3
10. 0.1875, or $\dfrac{3}{16}$, ohm

11 Rational equations and their graphs

11·1
1. 2
2. $\dfrac{1}{3}$
3. $\dfrac{a+7}{a+6}$
4. $\dfrac{2y}{y-9}$
5. $\dfrac{2}{y-8}$
6. $\dfrac{x+6}{x+4}$
7. $\dfrac{x+2}{5x-1}$
8. $\dfrac{x-4}{x-1}$
9. $\dfrac{x+5}{x-1}$
10. $\dfrac{a}{a-5}$

11·2
1. 1
2. $\dfrac{x^2+8x+15}{x+1}$
3. $\dfrac{7}{x}$
4. $\dfrac{3(a+5)(a-3)}{(a+4)(a-2)} = \dfrac{3a^2+6a-45}{a^2+2a-8}$
5. $(5-y)^2 = 25 - 10y + y^2$
6. $\dfrac{x-2}{x}$
7. $x - 9$
8. 2
9. $\dfrac{x+5}{2x}$
10. $\dfrac{x^2}{2}$

11·3

1. $\dfrac{9+x}{4}$

2. $\dfrac{1}{x+2}$

3. $\dfrac{3x}{x-4}$

4. $\dfrac{x-3}{(x-2)(x+1)} = \dfrac{x-3}{x^2-x-2}$

5. 2

6. $\dfrac{3x+27}{x+5}$

7. $\dfrac{3}{x+7}$

8. $\dfrac{(x-3)(x+1)}{x+4} = \dfrac{x^2-2x-3}{x+4}$

9. $\dfrac{x+2}{x-2}$

10. $\dfrac{(y+8)^2}{(y-8)^2}$

11·4

1. $\dfrac{5x-5}{6}$

2. $\dfrac{5x-20}{24}$

3. $\dfrac{6(x-1)}{(x+1)(x-2)} = \dfrac{6x-6}{x^2-x-2}$

4. $\dfrac{x^2+5x-15}{(x-5)(x+2)} = \dfrac{x^2+5x-15}{x^2-3x-10}$

5. $\dfrac{7x-4x^2}{4(4x-3)} = \dfrac{7x-4x^2}{16x-12}$

6. $\dfrac{5x-6}{(x+2)(x-2)} = \dfrac{5x-6}{x^2-4}$

7. $\dfrac{-5t}{(t-5)(t+5)} = \dfrac{-5t}{t^2-25}$

8. $\dfrac{15}{(x+3)(x-3)} = \dfrac{15}{x^2-9}$

9. $\dfrac{10x^2-3x-4}{3x(x-2)} = \dfrac{10x^2-3x-4}{3x^2-6x}$

10. $\dfrac{2x^2+3x+5}{(x+5)(x-5)} = \dfrac{2x^2+3x+5}{x^2-25}$

11. $\dfrac{a^2+6a+3}{(a+7)(a-3)^2}$

12. $\dfrac{x+14}{(x-3)(x-2)}$

13. $\dfrac{2x-3}{x-2}$

14. $\dfrac{x^3+3x^2-11x-1}{(x+5)(x-5)} = \dfrac{x^3+3x^2-11x-1}{x^2-25}$

15. $\dfrac{x+24}{(x+3)(x+2)(x-3)}$

11·5

1. $\dfrac{x^2+y^2}{x^2-y^2}$

2. $y+x$

3. $\dfrac{24}{1-4x}$

4. $\dfrac{x}{y}$

5. $\dfrac{-y^2}{xy-x^2+x-y}$

6. $\dfrac{1-x}{1+x}$

7. $\dfrac{x-5}{x+3}$

8. $\dfrac{x-4}{x+4}$

9. $\dfrac{y(4+y)}{2+3y} = \dfrac{4y+y^2}{2+3y}$

10. $\dfrac{3x+4}{x+1}$

11·6

1. $x = \dfrac{4}{3}$

2. $x = 3$

3. $x = 24$

4. $a = 8$

5. $x = \dfrac{1}{21}$

6. $x = 10$

7. $x = 5$

8. $\cancel{x=0}$, $x = 3$

9. $t = 2$

10. $x = 5$

11·7

1. $x = \dfrac{1}{3}, x = 4$

2. $x = -0.9$

3. $x = 6$

4. $x = \dfrac{5}{3}$

5. $x = \dfrac{2}{3}$

6. $y = -5$

7. $x = 3$

8. $x = 8$

9. $x = -9$

10. $x = 4$

11·8

1.

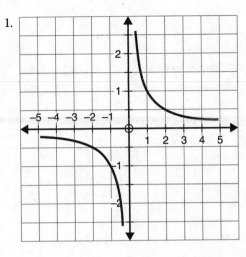

2.

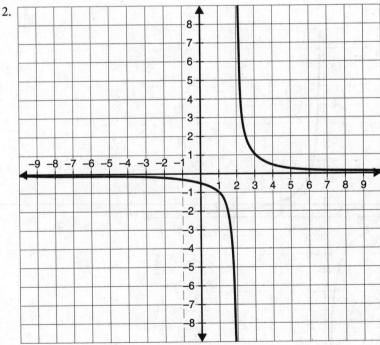

3.

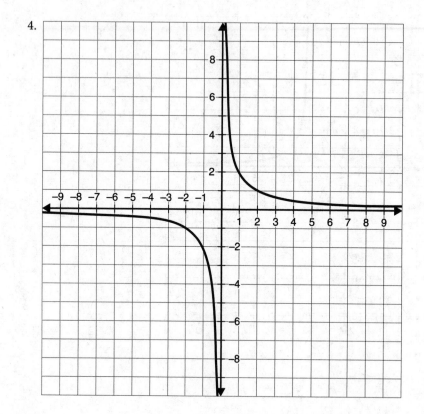

4.

5.

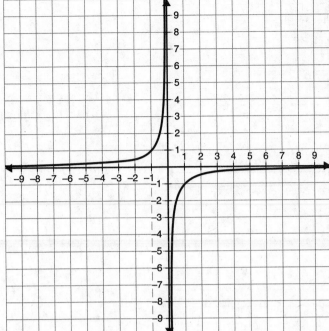

6.

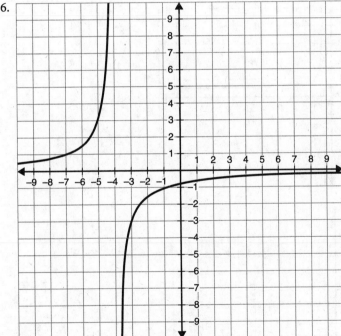

7.

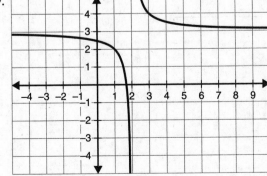

8.

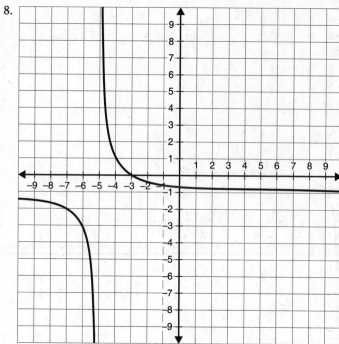

9.

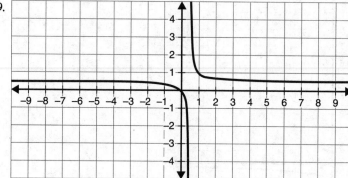

10.

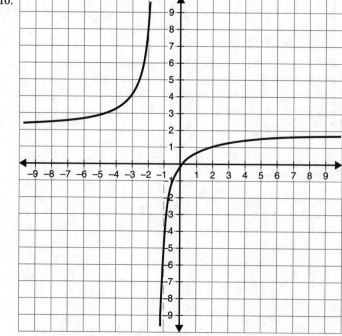

12 Exponential growth and decay

12·1
1. Exponential
2. Linear
3. Quadratic
4. Exponential
5. Quadratic

6. $a = 2, b = 3$
7. $a = 3, b = 2$
8. $a = \frac{1}{2}, b = 3$
9. $a = -2, b = \frac{1}{3}$
10. $a = 2, b = 10$

12·2
1. $y = 162$
2. $y = 8$
3. $y = -125$
4. $y = 9$
5. $y = 3,000,000$

6. $y = -0.02$
7. $x = 2$
8. $x = 2$
9. $x = 3$
10. $x = 4$

12·3
1. $5304.50
2. $14802.44
3. $2977.73
4. $3052.24
5. $327,148.96

6. $6204.59
7. $8492.03
8. $5977.69
9. $30,491.91
10. $59,913.95

12·4
1. Growth
2. Decay
3. Decay
4. Growth
5. Decay

6. Growth, 3,355,443,200 bacteria
7. Decay, 102.4 mg
8. Growth, approximately 379,518
9. Decay, $16,355.33
10. Decay, 36,652.78 acres

12·5
1.

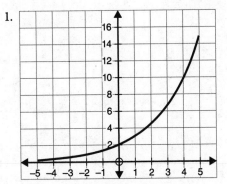

2.

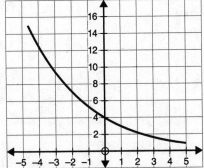

3.

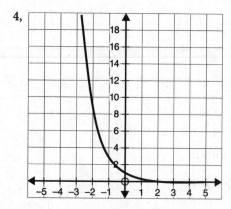

4,

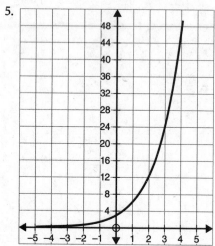

5.

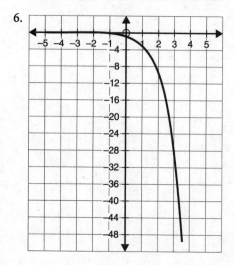

6.

7.

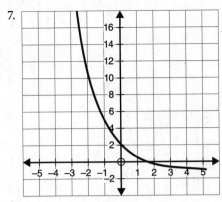

8.

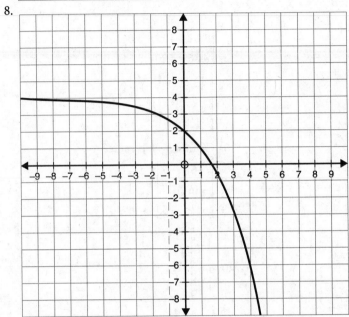

9.

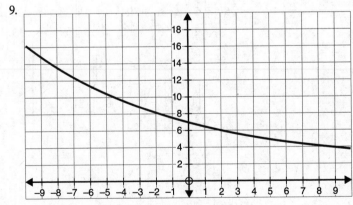

10.

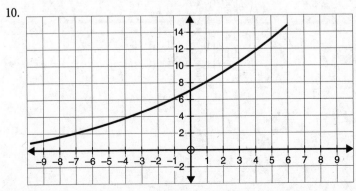

13 Matrix algebra

13·1
1. 2×4
2. 1×4
3. 3×2

4. 2×1
5. 3×3

	Tennis	Golf	Volleyball	Softball	Basketball
Equipment	24	15	2	7	3
Clothing	5	2	0	2	1
Accessories	0	3	1	5	0
Books	2	12	0	0	1

6. (matrix shown above)

13·2

1. $\begin{bmatrix} 6 & 11 \\ 7 & 10 \end{bmatrix}$

2. $\begin{bmatrix} 1 & 3 & 15 \\ 9 & 1 & -3 \end{bmatrix}$

3. $[10 \quad 6 \quad 8 \quad 3 \quad 2 \quad 10 \quad 3]$

4. Cannot be added

5. $\begin{bmatrix} \frac{4}{5} \\ \frac{5}{6} \end{bmatrix}$

6. $\begin{bmatrix} 1.01 & 4.12 & 3.04 & 2.12 \\ 4.36 & 4.20 & 3.54 & 0.28 \end{bmatrix}$

7. $\begin{bmatrix} 8 & 5 & 4 \\ 5 & 3 & 3 \\ 3 & 8 & 8 \end{bmatrix}$

8. Cannot be subtracted

9. $\begin{bmatrix} -6 & -5 \\ -16 & 2 \\ 1 & -6 \end{bmatrix}$

10. $\begin{bmatrix} \frac{1}{2} \\ \frac{2}{3} \end{bmatrix}$

13·3

1. $\begin{bmatrix} -15 & 25 \\ 35 & -5 \end{bmatrix}$

2. $\begin{bmatrix} -5 & -15 & 10 & -16 \end{bmatrix}$

3. $\begin{bmatrix} 21 & 18 & 39 \\ 47 & 7 & 8.5 \end{bmatrix}$

4. $\begin{bmatrix} -4.8 & -0.6 \\ 3.6 & -2.4 \\ 0 & 6.0 \end{bmatrix}$

5. $\begin{bmatrix} 40 \\ 45 \\ 52.5 \\ 20 \end{bmatrix}$

6. $3\begin{bmatrix} 1 & 3 & 4 \\ 5 & 2 & 6 \\ 2 & 1 & 3 \end{bmatrix} + 2\begin{bmatrix} 2 & 6 & 3 \\ 4 & 5 & 1 \\ 0 & 6 & 3 \end{bmatrix}\begin{bmatrix} 7 & 21 & 18 \\ 23 & 16 & 20 \\ 6 & 15 & 15 \end{bmatrix}$

7. $-2\begin{bmatrix} 3 & 2 \\ 5 & 1 \\ 0 & 3 \end{bmatrix} + 7\begin{bmatrix} 2 & 5 \\ -1 & -2 \\ 3 & 0 \end{bmatrix}\begin{bmatrix} 8 & 31 \\ -17 & -16 \\ 21 & -6 \end{bmatrix}$

8. Cannot be added

9. $\begin{bmatrix} -9 \\ -6 \\ -6 \end{bmatrix}$

10. $\begin{bmatrix} 0 & 0 & 0 & 0 & 0 \\ 0 & 0 & 0 & 0 & 0 \end{bmatrix}$

11.

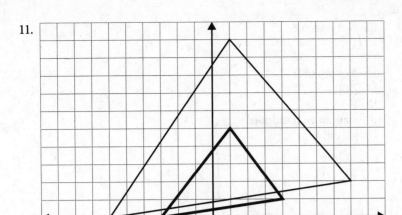

12.

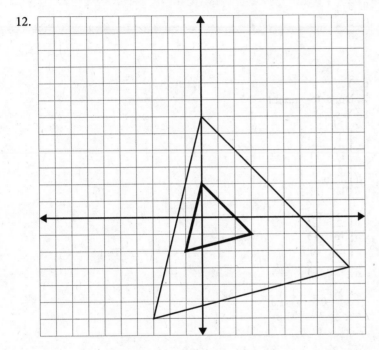

13.

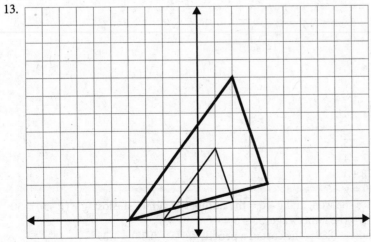

14. $\begin{bmatrix} \text{Senior} & \text{Junior} & \text{Sophomore} & \text{Freshman} \\ \$2234.28 & \$1143.12 & \$1376.94 & \$545.58 \end{bmatrix}$

13·4
1. 2×3
2. Not possible
3. Not possible
4. 3×2
5. 2×2
6. 3×3
7. Not possible
8. Not possible

9. 2×1

10. Not possible
11. 1×1
12. 3×3
13. Not possible
14. $[10]$
15. $[-5]$
16. Cannot be multiplied
17. $[48 \ \ 9]$

18. $\begin{bmatrix} 10 & 6 & 2 \\ 2 & 4 & -22 \end{bmatrix}$

13·5
1. -4
2. -22
3. Not possible
4. 71
5. Not possible
6. 1

7. 7
8. 0
9. 3
10. 5
11. 0

13·6
1. Inverses
2. Not inverses
3. Not inverses

4. Yes

5. No, not square

6. Yes
7. No, determinant $= 0$
8. Yes

9. $\begin{bmatrix} \dfrac{1}{11} & \dfrac{-3}{11} \\ \dfrac{2}{11} & \dfrac{5}{11} \end{bmatrix}$

10. $\begin{bmatrix} \dfrac{2}{13} & \dfrac{-1}{13} \\ \dfrac{-5}{13} & \dfrac{9}{13} \end{bmatrix}$

13·7
1. $x = 6, \ y = -1$
2. $x = 4, \ y = 1$
3. $x = -0.5, \ y = 1.3$

13·8
1. $x = 6, y = -1$
2. $x = 4, y = 1$
3. $x = -0.5, y = 1.3$
4. $x = 3, y = 1$
5. $x = 7, y = -3$
6. $x = -5, y = 2$

14   Problem solving

14·1
1. Let $N =$ the number. $N - 9 = 75$
2. Let $x =$ the number. $3x - 17 = 43$
3. Let $n =$ the number. $5x = 28 + x$
4. Let $y =$ the number. $8y - 40 = y - 5$
5. Let $g =$ the number of games she won, and $g - 6 =$ the number of games she lost. $g + (g - 6) = 30$
6. Let $S =$ the son's age, and $5S =$ the father's age. $5S - S = 44$
7. Let $x =$ the cost of a pen and $x - 0.89 =$ the cost of a pencil. $x + (x - 0.89) = 1.25$
8. Let $J =$ the number of laps Jaden ran and $J + 3 =$ the number of laps Carlos ran. $\frac{1}{4}(J + (J + 3)) = 6.25$
9. Let $P =$ the price of a large bucket of popcorn and $2P =$ the price of a ticket. $2(2P) + P = 35$
10. Let $W =$ the number of weeks and $5W =$ the number of work days. $5.50(5W) + 8W = 110$

14·2
1. $x = 1\frac{1}{3}$ ounces of 60% cocoa
2. Width = 14.5 feet and length = 35.5 feet
3. $k = 8.5$, length = 27
4. The number is 60.
5. The numbers are 31, 32, and 33.
6. 30 mg of full strength and 70 mg of 50% strength.
7. The numbers are 88, 89, 90, 91, and 92.
8. The numbers are 28, 30, 32, and 34.
9. An increase of 11.1% would be required.
10. $k = 9$, the height is 12 cm.

14·3
1. $x = 5$ or $x = 17$
2. $x = \sqrt{111} \approx 10.5$
3. 3, 5, and 7 or −1, 1, and 3
4. The base is 6 meters.
5. Approximately 8.1 seconds.
6. The dimensions of the rectangle are 6 inches by 8 inches.
7. Approximately 1.7 seconds
8. The original rectangle was 2 feet by 6 feet or 3 feet by 9 feet.
9. The two numbers are 7 and −2.
10. Approximately 1.2 seconds

14·4
1. 9 hours
2. 3 months
3. 2.1 days
4. $13\frac{1}{3}$ hours
5. 12 hours
6. 67.5 minutes
7. 3 cm
8. 5.5 hours
9. 2.6×10^{12} Newtons
10. 3 feet

14·5
1. 7 nickels
2. 1.25 hours
3. $2,600 at 4% and $3,400 at 7%
4. 12 problems correct
5. First number is 16, second is 10.
6. 10 pounds of $1.60 per pound tea and 30 pounds of $2 tea
7. 9 web pages
8. 188 chickens
9. 36 brownies
10. There are a total of 20 coins, 13 nickels, and 7 dimes.

Notes

Notes

Notes

Notes

Notes